**INTERNATIONAL INDIAN OCEAN EXPEDITION**
**METEOROLOGICAL MONOGRAPHS**     *Number 5*

The International Indian Ocean Expedition Meteorological Monographs, published by the East-West Center Press, contain detailed discussions and supporting data on the various components of the general atmospheric circulation over the Indian Ocean, as well as the results from measurements of atmosphere-ocean interaction made as part of the expedition's observational program. Manuscripts are solicited, and should be sent to C. S. Ramage, Department of Geosciences, University of Hawaii, Honolulu, Hawaii, U.S.A. 96822.

*Editorial committee*
C. S. RAMAGE     M. A. ESTOQUE     MICHAEL GARSTANG

*An Investigation of Heat Exchange*

# AN INVESTIGATION OF HEAT EXCHANGE

*by Donald J. Portman
and Edward Ryznar*

Honolulu      EAST-WEST CENTER PRESS

*The publication of this volume has been aided by Grant No. GA–386
from the National Science Foundation.*

# *Acknowledgment*

The investigation described in this monograph was conducted under U.S. National Science Foundation Grant Number G–22388. Its primary purpose was to establish a measurement network to help determine the heat and water vapor exchange at the air-sea interface for the International Indian Ocean Expedition (IIOE). The investigation was to fulfill part of the United States meteorological program recommended by the National Academy of Science Working Group on Meteorology for the IIOE and to meet, in part, one of the major objectives of the meteorological phase of the IIOE as expressed by the Scientific Committee on Oceanic Research of the International Council of Scientific Unions. In keeping with the general philosophy of the IIOE, it was desired that the measurement program contribute to related scientific activities of the host stations as well as provide basic data for the atmospheric and oceanographic sciences. The major emphasis was placed on determination of space and time distributions of solar and atmospheric radiation received at the ocean's surface. The information was meant to supplement radiation measurements made from artificial earth satellites.

The authors wish to express their appreciation to the many people whose work contributed to the investigation. Special thanks go to Mrs. Lenwood Paddock for efficiently handling the seemingly endless communications and many administrative details involved in travel and shipping arrangements. She was also responsible for preparing the tables and typing the manuscript for this monograph. Messrs. Hung Kwong Soo, Kenneth P. MacKay, Jr., and Arifhusen Waqif were responsible for the installation of the recording stations. They made all local arrangements, installed the equipment, made the necessary calibrations, and instructed local personnel in maintenance and operational details. Dr. Vincent Noble and Messrs. Jack Casey and James Kroth assisted in the design and testing of modifications of sensors and recording systems. Mr. Ramana K. Rao was responsible for the computations of precipitable water given in Appendix A.

The cooperation of seven different national meteorological services was required to establish fourteen radiation

*Acknowledgment (continued)*

recording stations. Personnel of these services gave generously of their time and shared in the responsibility of carrying out the investigation. Their contributions, ranging from administrative decisions and directives to digging ditches for electrical conduits, are gratefully acknowledged. Without such cooperation, of course, the investigation could not have been conducted.

Through the efforts of Brig. Gen. Benjamin Holzman, Ret., then director of Air Force Research Laboratories, we were able to send all radiation recording equipment directly by air from the United States to New Delhi via the U.S. Military Air Transport Service. The savings in time and money thus achieved were significant to the success of the project.

Professor Colin S. Ramage, director of the IIOE Scientific Program for Meteorology, and his staff were most helpful in planning and coordinating activities. Cloudiness information and sea-surface temperature data used in the analysis were provided by his office. The editorial help of Mrs. Ethel McAfee is gratefully acknowledged.

Finally, we would like to acknowledge the very important assistance provided by The British Phosphate Commission personnel in establishing and maintaining the recording station on Christmas Island. Mr. Ng Bah Kiat of that organization is to be especially commended for his devotion to maintaining the recording system there in reliable working order.

Donald J. Portman (1), University of Michigan
Edward Ryznar (2), University of Michigan

Ann Arbor, 28 August 1969

# Contents

# *Abstract*

Island and coastal recording stations were established mainly in India and the western half of the Indian Ocean to measure incident solar radiation and atmospheric radiation for the International Indian Ocean Expedition. Total incident solar radiation data were obtained at several stations from 1963 through 1965 and are analyzed to determine (1) the variation of daily totals for cloudless days and for actual cloudiness throughout the year for certain stations, (2) a latitudinal distribution of daily totals for cloudless days throughout the year, and (3) distributions for the Indian Ocean for each month of 1963 and 1964 using observed cloudiness data averaged for 5-degree latitude-longitude squares. Distributions of net radiative exchange for the same months were computed by using information obtained from the following: total incident solar radiation data; and computations of reflected solar radiation, emitted long-wave radiation, and an average value of atmospheric radiation.

The distributions of incident solar radiation and net radiative exchange are similar to those obtained by Budyko (*3*) and Mani *et al.* (*4*) for the Indian Ocean, but values are smaller. Two possible reasons for smaller values of incident solar radiation are discussed. One is that there was an abnormally large amount of volcanic material in the atmosphere from the eruption of Mt. Agung, Bali, in March 1963. The other is that the data used in this study may have been more nearly representative of the atmosphere over the ocean than were the data used for the earlier studies.

Equipment and procedures are described, and the results of the analysis are presented in both tabular and graphical form. Appendix A describes the method used to compute precipitable water amounts for clear day solar radiation calculations. Appendix B consists of tabulated incident solar and total hemispherical radiation data.

# PART 1: *Introduction*

## I. BACKGROUND

The seasonal and geographical distributions of variables comprising the energy exchange over the oceans of the earth have been the subject of a number of investigations over a period of many years. The applicability of the results of the investigations to dynamic problems of the atmosphere and oceans was limited, however, because of basic restrictions in both space and time resolution. A need for more detailed measurements of the components of radiative and turbulent exchange for the oceans was evident. During the International Geophysical Year (IGY) and the second half of the 1950's, the number of stations for measuring thermal radiation increased significantly throughout the world. The new data provided by the increased station network and also by a number of research expeditions included measurements for areas for which previously there were very few data. Budyko (*3*) used the new observations and improved existing methods of computation to compile more accurate and detailed maps of world radiative exchange.

An opportunity to increase observational data specifically for the Indian Ocean was provided by the International Indian Ocean Expedition (IIOE), which had as one of its main objectives a program to observe and describe and, if possible, explain the circulations of the Indian Ocean and the atmosphere above it and the interaction between them. Measurements of the components of radiative exchange were an integral part of the proposed meteorological observation program. The following components were identified for a meaningful delineation of the radiative exchange at the air-sea interface:

1. Incident solar radiation (direct plus diffuse)
2. Atmospheric radiation
3. Reflected solar radiation
4. Radiation emitted by the water surface

Personnel of the Department of Meteorology and Oceanography, University of Michigan, participated in the IIOE by carrying out a measurement program designed to provide basic information on items 1 and 2. For that purpose, a number of radiation recording stations was established on island and coastal areas at locations where electrical power was available (or was to be available) and where weather service or communication technicians could assist in the installation and operation of the equipment.

In the absence of clouds, measurements of incident solar and atmospheric radiation made on small islands and on windward coasts may be expected to represent open-ocean locations. Measurements made with cloudiness, on the other hand, cannot so easily be extrapolated. It was hoped that satellite cloud photography would assist in data evaluation in this respect.

Reflected solar radiation and radiation emitted by the water surface (items 3 and 4, above) were not measured directly. Estimates of their seasonal and geographical distributions were determined for the study presented here with the aid of empirical relationships for reflected solar radiation and of maps of average sea-surface temperature for emitted radiation. These results are combined with the measurements to yield monthly averages of the geographical distributions of net radiation exchange for 1963 and 1964.

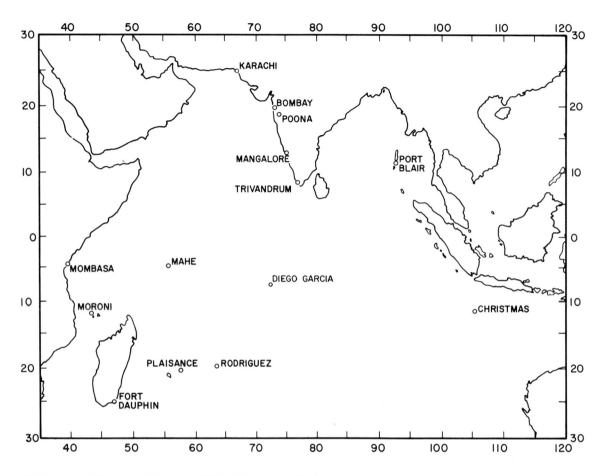

FIGURE 1: Locations of island and coastal stations for recording thermal radiation.

## II. STATIONS AND AGENCIES

Fourteen island and coastal thermal-radiation recording stations were installed in the period March through December 1963. Table 1 lists each station, its location, and the responsible meteorological agency.

The location of each of the fourteen stations is shown on the map in Figure 1. Except for Christmas Island and Port Blair, the stations were located in the western half of the Indian Ocean between about 40E and 80E. The northernmost station was Karachi at 25N, and the southernmost was Fort Dauphin at 25S. All stations were less than 300 feet above mean sea level.

## III. MEASUREMENTS AND SENSORS

Continuous radiation measurements at each of the stations consisted of (1) total incoming solar radiation (direct plus diffuse) and (2) total incoming solar *and* atmospheric radiation. Atmospheric radiation is obtainable by subtracting the former from the latter.

### A. Total Solar Radiation

Eppley (180°) pyranometers (pyrheliometers) were used to measure solar radiation. An instrument is illustrated in Figure 2 (5). The sensing element, enclosed in a soda-lime glass bulb, consists of two concentric rings of equal sur-

**Table 1: Island and Coastal Stations for Recording Thermal Radiation**

| Station | Location | | Meteorological Agency |
|---------|----------|---|----------------------|
| Karachi, West Pakistan | 25N | 67E | Pakistan Meteorological Department |
| Bombay, India | 19N | 73E | India Meteorological Department |
| Poona, India | 18N | 74E | India Meteorological Department |
| Mangalore, India | 13N | 75E | India Meteorological Department |
| Port Blair, South Andaman Island, Andaman Islands | 12N | 93E | India Meteorological Department |
| Trivandrum, India | 8N | 77E | India Meteorological Department |
| Mombasa, Kenya | 4S | 39E | East African Meteorological Department |
| Port Victoria, Mahé Island, Seychelles Islands | 5S | 55E | East African Meteorological Department |
| Diego Garcia Island, Chagos Islands | 8S | 72E | Mauritius Meteorological Service |
| Christmas Island | 11S | 106E | Australian Bureau of Meteorology |
| Moroni, Grande Comore Island, Comores Islands | 13S | 45E | Service Météorologique des Comores |
| Rodriguez Island, Mascareign Islands | 19S | 63E | Mauritius Meteorological Service |
| Plaisance, Mauritius Island, Mascareign Islands | 20S | 57E | Mauritius Meteorological Service |
| Fort Dauphin, Madagascar | 25S | 47E | Météorologie Nationále, Republika Malagasy |

FIGURE 2: The Eppley (180°) pyranometer (pyrheliometer).

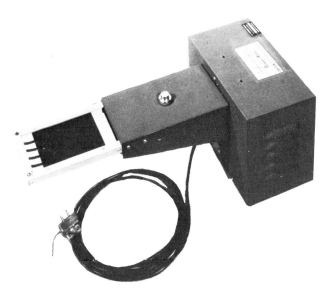

FIGURE 3: Modified Beckman and Whitley total hemispherical radiometer.

face area, one white and one black. The temperature difference between the rings is measured by means of ten thermojunctions in thermal contact with the lower surfaces of the rings. Connected to the thermopile is a temperature compensation circuit designed to maintain a linear relationship between radiation received and voltage produced throughout a temperature range of 0° C to 50° C.

### B. Total Hemispherical Radiation

Modified Beckman and Whitley total hemispherical radiometers were used to measure the total radiation from the sun and atmosphere. An instrument is illustrated in Figure 3. The sensing element is a flat plate that contains a 720-junction silver-constantan thermopile arranged so that its voltage output is proportional to the heat conducted perpendicularly through the plate. The upper surface of the plate is coated with Parsons Optical Matte Black Lacquer; the lower surface is polished aluminum. To minimize variations of convective heat transfer from it, the plate is mounted in a forced air stream supplied by an electric blower housed in the main support of the instrument. An aluminum radiation shield, 0.5 inch below the plate, has a black upper surface and a polished lower one.

Analyses of the energy balance of the sensing element and its shield have been made by Gier and Dunkle (6) and by Portman and Dias (7). It can be shown that, if the calibration factor is known, the total incident thermal radiation can be determined from measurements of the thermopile output and the upward-directed radiation of the black receiving surface. To correct for temperature dependency of the thermopile and to measure radiation emitted by the sensor's upper surface, compensation and radiation analog circuits were designed and tested (MacKay, 8). The circuits were adapted for manufacture by Beckman and Whitley, Inc., in a special order for the fifteen instruments required for the Indian Ocean investigation.

The basic relationship of the thermopile, its temperature compensation circuit, and the analog circuit for emitted radiation are shown in a block diagram in Figure 4. These components simulate the radiometer equation and produce a voltage, $V_{out}$, which is proportional to total incident radiation. The thermopile output, $V_T$, depends on both the temperature of the sensing plate and incident radiation. The compensation circuit, by means of a thermistor, alters the output of the thermopile in accordance with the calibration curve supplied by the manufacturer, shown as $V_T f(T)$. The analog circuit, on the other hand, generates a voltage equivalent to the radiation emitted by the plate, $e \, \sigma \, T^4$ ($e$ for the plate was close to unity). The analog voltage was added to the compensated voltage, and the sum of the two, $V_{out}$, is measured.

### IV. RECORDER-INTEGRATOR SYSTEM

Special twin recorder-integrator systems were designed and manufactured by Minneapolis Honeywell, Inc., for recording both the instantaneous and the time integral value of the two radiation sensor outputs at each station. A block diagram of the system is shown in Figure 5. A complete recording system consisted of the following components: one constant voltage transformer, two operational amplifiers, two integrators, and two Honeywell Electronik 17 strip chart recorders (Fig. 6). The chart speed of the recorder was 2 inches an hour; a chart roll lasted about a month.

The operation of the integrator pen is controlled by the position of the analog pen through an electromechanical linkage involving a potentiometric slide wire, an amplifier, a motor, a gear train, and a cam. The analog pen positions a slide wire so that a specific voltage is produced, amplified, and supplied to the motor. In this way, the motor speed is proportional to the scale position of the analog pen. By means of a gear train, then, a cam is made to revolve once for each input of 10 langleys. (A langley is 1 gram-calorie per cm² and is abbreviated as ly.) Owing to nonlinearities in its starting characteristics, the electric motor is kept running at a low speed so that the integrator indicates one traverse (10 ly) every 20 minutes with no output from the sensor. The system was designed to operate on 50-Hz power frequency.

### V. INSTALLATION AND MAINTENANCE OF EQUIPMENT

The installation of equipment at a station consisted of (1) selecting a site that had suitable exposure and electrical power for the radiation sensors, (2) preparing supports for and mounting and leveling the sensors, (3) calibrating the recorder-integrator system, and (4) instructing station personnel in the maintenance of equipment. The sensors were mounted near ground level wherever possible to minimize errors brought about by wind effects on the radiometer sensing plate. Station personnel were provided with instruction manuals and were instructed in duties such as cleaning sensors, checking calibrations, and maintaining the recorder ink supply.

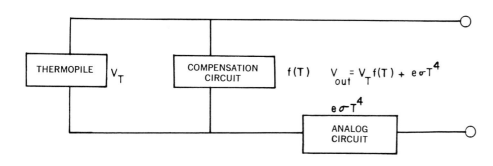

**FIGURE 4: Block diagram of radiometer circuitry.**

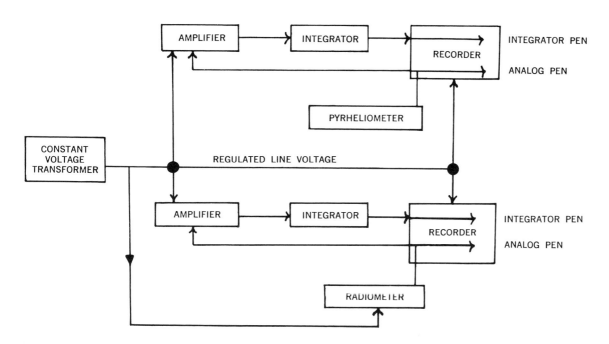

**FIGURE 5: Block diagram of recorder-integrator system.**

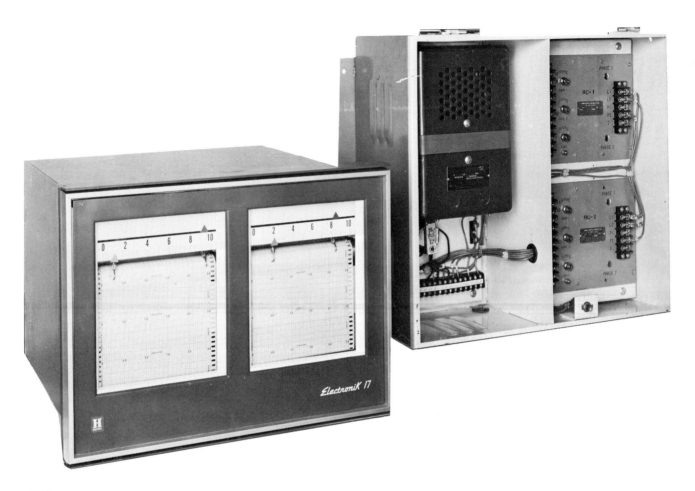

**FIGURE 6:** Recording system for thermal radiation.

**FIGURE 7:** Sensor installations
   (A) Karachi, West Pakistan
   (B) Bombay, India
   (C) Trivandrum, India
   (D) Mombasa, Kenya
   (E) Diego Garcia Island
   (F) Christmas Island
   (G) Moroni, Comores Islands
   (H) Rodriguez Island
   (I) Plaisance, Mauritius Island
   (J) Fort Dauphin, Madagascar

## VI. DESCRIPTION OF STATIONS

### A. Karachi, West Pakistan (Fig. 7A)

The radiation sensors were installed on top of a wall extending above the roof of the Meteorological Department's Institute of Meteorology and Geophysics in Nazimabad. The recording system was housed on the second floor of the Institute building. Installation was completed on 17 March 1963.

### B. Bombay, India (Fig. 7B)

The sensors were installed approximately 300 feet from the radiosonde station at Santa Cruz Airport, about 2 miles inland. They were mounted about 5 feet above the ground on a "T" stand with a concrete base. The recorders were placed in the radiosonde building. Waterproof cables connecting the sensors and the recording equipment were buried about 18 inches below the ground. Installation was completed on 20 March 1963.

### C. Poona, India

The sensors were installed on the roof of the Meteorological Department building near similar radiation sensors. The recording units were placed on the second floor of the same building. Installation was completed in August 1963.

### D. Mangalore, India

The sensors were installed near existing solarimeters on the roof of a storage shed near the Meteorological Department Observatory, about 6 miles inland. The recording equipment was housed inside the shed. Installation was completed on 18 April 1964.

### E. Port Blair, South Andaman Island, Andaman Islands

The sensors were installed on the roof of the Meteorological Department Observatory, and the recording equipment was housed inside the Observatory. The Observatory was located on a peninsula about 1 mile wide. Installation was completed on 15 December 1963.

Karachi, West Pakistan

Bombay, India

**F. Trivandrum, India (Fig. 7C)**

The radiation sensors were installed near an observatory of the Meteorological Department, located about 2 miles inland. The recording equipment was located in the observatory itself about 80 feet from the sensors. Installation was completed on 15 April 1964.

**G. Mombasa, Kenya (Fig. 7D)**

The sensors were mounted on a theodolite stand about 250 feet from the East African Meteorological Department Observatory, Mombasa Airport, about 6 miles inland. The recording system was located in the observing station. Installation was completed on 12 May 1963.

Trivandrum, India

### H. Port Victoria, Mahé Island, Seychelles Islands

The sensors were mounted on a wooden platform on the roof of the East African Meteorological Station at Long Pier, Victoria. The recording equipment was located inside the station building. Installation was completed on 24 June 1963.

### I. Diego Garcia Island, Chagos Islands (Fig. 7*E*)

The system for Diego Garcia Island was installed by the end of July 1963 by personnel of the Mauritius Meteorological Service. Difficulties with electrical power and equipment, however, prevented recordings from being obtained.

Mombasa, Kenya

Diego Garcia Island

**Christmas Island**

**Moroni, Comores Islands**

### J. Christmas Island (Fig. 7F)

The sensors were mounted about 3 feet above ground in an open field approximately 250 feet from the radio station. The recording system was located in the station. Installation was completed in mid-August 1963.

### K. Moroni, Grande Comore Island, Comores Islands (Fig. 7G)

The sensors were mounted on the roof of the headquarters of the Service Météorologique des Comores, about 0.5 mile inland. The recorders were housed inside the same building. Installation was completed on 10 May 1963. Because power was usually turned on only for 0.5-day intervals, complete records were not available until 24-hour power began in March 1964.

### L. Rodriguez Island, Mascareign Islands (Fig. 7H)

The sensors were installed on the roof of the local weather observatory, about 500 feet inland. The recording unit was installed in the observatory building. The system was installed in May 1963, but it did not operate satisfactorily because of intermittent electrical power and frequent failure of a d-c to a-c voltage inverter unit.

### M. Plaisance, Mauritius Island, Mascareign Islands (Fig. 7I)

The sensors were mounted about 4 feet above the ground within a compound for meteorological instruments of the Mauritius Meteorological Service, about 3 miles inland. The recorders were inside the observatory. Installation was completed on 26 May 1963.

### N. Fort Dauphin, Madagascar (Fig. 7J)

The sensors were mounted on a small platform about 1 foot above the roof of the old terminal building at Fort Dauphin Airport, about 1 mile inland. The recording system was housed in the terminal building near the observing station of the Direction de la Météorologie Nationale. Installation was completed on 19 April 1963.

Rodriguez Island

Plaisance, Mauritius Island

Fort Dauphin, Madagascar

# PART 2: *Data and Discussion*

## VII. DATA INVENTORY

Data from the radiation recording stations have been received and processed at the University of Michigan and also at several meteorological department centers under whose authority the stations were established. Most of the data received at the University of Michigan were processed; they are tabulated in Appendix B in the form of daily sums. These data were received in the form of chart rolls (two rolls from each station a month) and were immediately inspected to ascertain whether the recording systems were functioning properly. If the recordings showed no obvious deficiencies, a simple visual-manual method of abstracting hourly averages was used. The hourly averages were totaled for each 24-hour day, beginning with the hour ending 0100 local time; the number of langleys a day were then computed and tabulated.

The status of other data collected in 1963, 1964, and 1965 is shown in Table 2. The number of days of recordings for each month obtained for each sensor is listed in Table 2. The processed data are tabulated in Appendix B. The letter symbols used in Table 2 have the following meanings:

A: *Original data, when recorded, held by local meteorological departments.* Because all equipment was transferred to local meteorological departments in early 1966, in effect, almost all data collected after 1965 come under this category. In many cases radiometers had deteriorated to such an extent that they were unreliable prior to the transfer. These and other equipment difficulties are likely to have forced recordings to be terminated in many locations unless replacement components were obtained.

B: *Data processed by the Indian Meteorological Department and provided to the University of Michigan.* Copies are available on request.

C: *Data for part days only, held at the University of Michigan and available on request.* In this category are data from the station at Moroni, where, until April 1964, electrical power was available only during the morning hours. The data have been reduced to hourly averages except for total hemispherical radiation data obtained in 1965 indicated by the letter $C_1$.

D: *Unreliable data or no data because of equipment malfunction or failure.* A major difficulty during the first year of recording was the malfunction and failure of electronic amplifiers in the recorders. After it was learned that the trouble stemmed from faulty components mistakenly used during factory production, new units were supplied, and subsequently all recorders received new amplifiers. If erroneous data resulting from equipment malfunction were found, remedial action was taken as soon as possible. Chart transit time to the University of Michigan was usually one month. Because of this time lag between data recording and evaluation, at least one month of erroneous data would have been recorded.

E: *No data because of inadequate or unavailable electrical power.* When it became apparent that electrical power could not be made available for the stations at Rodriguez and Diego Garcia, the equipment was transferred by the Mauritius Meteorological Department to Mauritius.

## VIII. ANALYSIS AND DISCUSSION

### A. Incident Solar Radiation for Cloudless Days

The starting point in the analysis of radiation measurements was the determination of the annual variation of daily totals of incident solar radiation for cloudless days. In this context a cloudless day is defined as one whose pyranometer record showed a uniformly increasing and then decreasing pattern, symmetrical around local noon, during the course of the day. Experience in the mid-latitudes, with instruments of comparable sensitivities, suggests that overcast clouds are seldom, if ever, thin enough and uniform enough through the course of a day, to escape detection by careful inspection of a pyranometer record.* In order to obtain additional data for stations that had very few completely cloudless days, data from days that had occasionally cloudy periods were used if the cloudy periods were short enough so that "cloudless" solar radiation for the periods could be estimated with some confidence.

Daily totals of incident solar radiation were obtained for all cloudless days (as defined above) from the time each station was placed into operation until December 1965. These data are given in Table 3. The total number of cloudless days was as follows: (1) for the Northern Hemisphere, Mangalore, 80; Port Blair, 48; Trivandrum, 42; (2) for the Southern Hemisphere, Mombasa, 36; Mahé Island, 35; Christmas Island, 37; Plaisance, 61; and Fort Dauphin, 141. The annual variation of daily totals for each station is shown in Figure 8.

Curves *1* and *3* were obtained from the work of Bolsenga (*9*), who computed and tabulated daily totals of solar radiation for cloudless skies for the Northern Hemisphere from information pertaining to latitude, solar declination, precipitable water, and dust content. A solar constant of 1.98 ly min$^{-1}$ was used. The curves represent values that were taken from his tables and adjusted to account for solar distance and declination for the Southern Hemisphere. For Port Blair, Trivandrum, and Mahé Island, precipitable water values of 3 cm for the dry season and 4 cm for the rainy season were used on the basis of radiosonde analyses described below. For other stations, a precipitable water value of 4 cm was used for all months. Fractional dust depletion was taken to be 0.04 for all stations for all months.

---

* In some cases, the authors have noticed thin cirrostratus clouds only after their presence was indicated on a pyranometer record.

**Table 2: Data Inventory**

| | | | | 1963 | | |
|---|---|---|---|---|---|---|
| | | J | F | M | A | M |
| Bombay, India | Pyr | | | A | A | A |
| | Rad | | | A | A | A |
| Karachi, W. Pakistan | Pyr | | | 9 | 26 | 29 |
| | Rad | | A | A | | |
| Mombasa, Kenya | Pyr | | | | D | D |
| | Rad | | | | 19 | 7 |
| Fort Dauphin, Madagascar | Pyr | | | | 11 | 31 |
| | Rad | | | | 7 | 31 |
| Moroni, Comores Island | Pyr | | | | | C |
| | Rad | | | | | C |
| Rodriguez Island | Pyr | | | | | E |
| | Rad | | | | | E |
| Plaisance, Mauritius | Pyr | | | | | |
| | Rad | | | | | |
| Diego Garcia Island | Pyr | | | | | |
| | Rad | | | | | |
| Poona, India | Pyr | | | | | |
| | Rad | | | | | |
| Christmas Island | Pyr | | | | | |
| | Rad | | | | | |
| Mahé, Seychelles | Pyr | | | | | |
| | Rad | | | | | |
| Trivandrum, India | Pyr | | | | | |
| | Rad | | | | | |
| Mangalore, India | Pyr | | | | | |
| | Rad | | | | | |
| Port Blair, Andaman Islands | Pyr | | | | | |
| | Rad | | | | | |

| 1964 | | | | | | | | | | | | | 1965 | | | | | | | | | | | | | | | |
|---|---|---|---|---|---|---|---|---|---|---|---|---|---|---|---|---|---|---|---|---|---|---|---|---|---|---|---|---|
| A | S | O | N | D | J | F | M | A | M | J | J | A | S | O | N | D | J | F | M | A | M | J | J | A | S | O | N | D |
| A | A | A | A | A | A | A | A | A | A | A | A | A | A | A | A | A | A | A | A | A | A | A | A | A | A | A | A | A |
| A | A | A | A | A | A | A | A | A | A | A | A | A | A | A | A | A | A | A | A | A | A | A | A | A | A | A | A | A |
| A | A | A | A | A | A | A | A | A | A | A | A | A | A | A | A | A | A | A | A | A | A | A | A | A | A | A | A | A |
| A | A | A | A | A | A | A | A | A | A | A | A | A | A | A | A | A | A | A | A | A | A | A | A | A | A | A | A | A |
| 31 | 30 | 11 | D | D | D | 19 | D | D | D | D | D | D | D | D | D | D | D | D | D | D | 25 | 30 | 30 | 30 | 29 | 30 | 28 | 30 |
| D | D | D | D | D | 4 | 29 | 30 | 30 | 31 | 30 | 28 | 8 | D | 24 | 29 | 30 | 30 | 21 | 29 | 12 | D | D | D | D | D | D | D | D |
| 30 | 30 | 30 | 30 | 31 | 31 | 29 | 30 | 30 | 30 | 29 | 30 | 30 | 26 | 29 | 22 | D | D | D | D | D | 14 | 28 | D | D | D | 28 | 29 | 30 |
| D | D | D | D | D | D | D | D | D | D | D | D | D | D | 23 | 26 | 22 | 30 | D | 8 | D | D | D | D | D | D | D | D | D |
| C | C | C | C | C | C | C | C | C | D | D | D | D | D | D | D | D | D | D | D | D | A | A | A | A | A | A | A | A |
| C | C | C | C | C | C | C | C | C | C | D | D | D | D | D | D | D | D | D | D | D | D | $C_1$ | $C_1$ | $C_1$ | $C_1$ | $C_1$ | $C_1$ | D |
| E | E | E | E | E | E | E | E | E | E | E | E | E | C | C | E | E | E | E | E | E | E | E | E | E | E | E | E | E |
| E | E | E | E | E | E | E | E | E | E | E | E | E | C | C | E | E | E | E | E | E | E | E | E | E | E | E | E | E |
| 27 | 28 | 28 | 29 | 30 | 28 | 26 | 29 | 27 | 26 | 27 | 30 | 27 | 29 | 31 | 30 | 31 | D | D | D | 5 | 13 | 15 | 28 | 27 | 27 | 30 | 29 | 29 |
| 27 | 26 | D | D | D | D | 16 | 25 | 18 | D | D | D | 27 | 29 | 31 | 30 | 31 | D | D | D | D | D | 21 | 28 | 8 | D | D | D | D |
| E | E | E | E | E | E | E | E | E | E | E | E | E | E | E | E | E | E | E | E | E | E | E | E | E | E | E | E | E |
| E | E | E | E | E | E | E | E | E | E | E | E | E | E | E | E | E | E | E | E | E | E | E | E | E | E | E | E | E |
| A | A | A | A | A | A | A | A | A | A | A | A | A | A | A | A | A | A | A | A | A | A | A | A | A | A | A | A | A |
| A | A | A | A | A | A | A | A | A | A | A | A | A | A | A | A | A | A | A | A | A | A | A | A | A | A | A | A | A |
| 9 | 30 | 31 | 28 | 31 | 29 | 28 | 31 | 30 | 31 | 30 | 31 | 31 | 29 | 31 | 27 | 30 | 31 | 27 | 31 | 28 | 30 | 29 | 29 | 26 | 23 | 31 | 30 | 31 |
| 9 | 29 | 30 | 20 | 30 | 29 | 28 | 31 | 28 | 31 | 30 | 30 | 31 | 28 | 21 | 27 | 17 | 27 | 10 | D | D | D | D | 9 | D | D | D | D | D |
|  |  | 4 | 29 | 25 | 31 | 27 | 31 | 27 | 30 | 30 | 29 | 28 | 29 | 30 | 28 | 28 | 30 | 28 | 31 | 6 | A | A | A | A | A | A | A | A |
|  |  |  | D | D | D | D | D | D | D | D | D | D | D | D | D | D | D | D | D | D | A | A | A | A | A | A | A | A |
|  |  |  | B | B | B | B | B | B | B | B | B | B | A | A | A | A | A | A | A | A | A | A | A | A | A | A | A | A |
|  |  |  | A | A | A | A | A | A | A | A | A | A | A | A | A | A | A | A | A | A | A | A | A | A | A | A | A | A |
|  |  |  | B | B | B | B | B | B | B | B | B | B | B | B | B | A | A | A | A | A | A | A | A | A | A | A | A | A |
|  |  |  | A | A | A | A | A | A | A | A | A | A | A | A | A | A | A | A | A | A | A | A | A | A | A | A | A | A |
|  |  |  | B | B | B | B | B | B | B | B | B | B | B | B | B | A | A | A | A | A | A | A | A | A | A | A | A | A |
|  |  |  | A | A | A | A | A | A | A | A | A | A | A | A | A | A | A | A | A | A | A | A | A | A | A | A | A | A |

Table 3A: Total Incident Solar Radiation
for Cloudless Days
Mangalore, India (ly day$^{-1}$)

| 1963 | | 1964 | | | |
|---|---|---|---|---|---|
| 22 Dec. | 479 | 1 Jan. | 467 | 13 May | 633 |
| 28 | 467 | 2 | 461 | 15 | 627 |
| 29 | 464 | 3 | 458 | 28 | 602 |
| 30 | 467 | 4 | 461 | 29 | 613 |
| 31 | 467 | 5 | 478 | 28 July | 643 |
| | | 6 | 469 | 29 Aug. | 623 |
| | | 7 | 464 | 4 Sept. | 623 |
| | | 8 | 474 | 17 Oct. | 574 |
| | | 9 | 457 | 3 Nov. | 574 |
| | | 10 | 473 | 15 | 523 |
| | | 12 | 469 | 16 | 523 |
| | | 14 | 465 | 18 | 512 |
| | | 16 | 471 | 23 | 499 |
| | | 22 | 485 | 25 | 514 |
| | | 24 | 489 | 26 | 534 |
| | | 25 | 490 | 28 | 523 |
| | | 26 | 484 | 29 | 517 |
| | | 27 | 496 | 30 | 503 |
| | | 28 | 485 | 1 Dec. | 494 |
| | | 2 Feb. | 501 | 2 | 491 |
| | | 4 | 497 | 3 | 503 |
| | | 7 | 488 | 6 | 484 |
| | | 8 | 507 | 7 | 493 |
| | | 9 | 506 | 8 | 503 |
| | | 10 | 506 | 9 | 494 |
| | | 11 | 521 | 10 | 481 |
| | | 12 | 541 | 11 | 482 |
| | | 13 | 540 | 12 | 500 |
| | | 14 | 538 | 14 | 496 |
| | | 16 | 527 | | |
| | | 23 | 512 | | |
| | | 25 | 533 | | |
| | | 26 | 537 | | |
| | | 28 | 545 | | |
| | | 6 Mar. | 513 | | |
| | | 7 | 514 | | |
| | | 11 | 578 | | |
| | | 12 | 581 | | |
| | | 18 | 522 | | |
| | | 27 | 572 | | |
| | | 28 | 575 | | |
| | | 7 Apr. | 561 | | |
| | | 8 | 569 | | |
| | | 10 | 579 | | |
| | | 26 | 623 | | |

Table 3B: Total Incident Solar Radiation
for Cloudless Days
Port Blair, Andaman Islands (ly day$^{-1}$)

| 1963 | | 1964 | |
|---|---|---|---|
| 23 Dec. | 510 | 9 Jan. | 494 |
| 24 | 497 | 10 | 512 |
| | | 11 | 502 |
| | | 12 | 500 |
| | | 13 | 505 |
| | | 24 | 523 |
| | | 25 | 519 |
| | | 26 | 517 |
| | | 28 | 530 |
| | | 29 | 522 |
| | | 31 | 539 |
| | | 1 Feb. | 541 |
| | | 6 | 519 |
| | | 7 | 510 |
| | | 8 | 529 |
| | | 9 | 564 |
| | | 29 | 558 |
| | | 2 Mar. | 578 |
| | | 4 | 595 |
| | | 6 | 594 |
| | | 13 | 527 |
| | | 15 | 537 |
| | | 19 | 594 |
| | | 20 | 573 |
| | | 21 | 554 |
| | | 23 | 603 |
| | | 4 Apr. | 624 |
| | | 7 | 625 |
| | | 8 | 613 |
| | | 15 | 615 |
| | | 17 | 638 |
| | | 21 | 655 |
| | | 22 | 652 |
| | | 15 May | 662 |
| | | 2 June | 629 |
| | | 10 July | 654 |
| | | 25 | 636 |
| | | 8 Oct. | 596 |
| | | 20 | 577 |
| | | 10 Nov. | 524 |
| | | 17 | 517 |
| | | 25 | 498 |
| | | 5 Dec. | 497 |
| | | 11 | 497 |
| | | 13 | 482 |

Table 3C: Total Incident Solar Radiation
for Cloudless Days
Trivandrum, India (ly day⁻¹)

| 1963 | | 1964 | |
|------|-----|------|-----|
| 21 Nov. | 541 | 3 Jan. | 498 |
| 13 Dec. | 521 | 4 | 497 |
| 14 | 527 | 6 | 509 |
| 15 | 512 | 8 | 504 |
| 18 | 502 | 10 | 492 |
| 20 | 496 | 13 | 515 |
| 28 | 495 | 19 | 516 |
| 30 | 507 | 20 | 544 |
| | | 21 | 539 |
| | | 22 | 532 |
| | | 4 Feb. | 558 |
| | | 5 | 537 |
| | | 6 | 557 |
| | | 7 | 551 |
| | | 11 | 555 |
| | | 12 | 575 |
| | | 13 | 572 |
| | | 14 | 602 |
| | | 26 | 586 |
| | | 9 Mar. | 592 |
| | | 11 | 622 |
| | | 13 | 608 |
| | | 14 | 576 |
| | | 19 | 621 |
| | | 2 Apr. | 607 |
| | | 19 | 618 |
| | | 15 May | 598 |
| | | 16 | 607 |
| | | 28 | 579 |
| | | 20 June | 607 |
| | | 17 July | 614 |
| | | 17 Aug. | 620 |
| | | 21 | 629 |
| | | 25 | 631 |

Table 3D: Total Incident Solar Radiation
for Cloudless Days
Mombasa, Kenya (ly day⁻¹)

| 1963 | | 1964 | | 1965 | |
|------|-----|------|-----|------|-----|
| 7 June | 508 | 12 Feb. | 646 | 13 May | 547 |
| 23 Sept. | 611 | 13 | 642 | 20 | 528 |
| | | 15 | 656 | 1 June | 505 |
| | | 19 | 637 | 17 | 503 |
| | | 20 | 623 | 4 July | 504 |
| | | | | 14 | 526 |
| | | | | 22 | 554 |
| | | | | 2 Aug. | 532 |
| | | | | 19 | 525 |
| | | | | 31 | 568 |
| | | | | 4 Sept. | 613 |
| | | | | 5 | 593 |
| | | | | 18 | 628 |
| | | | | 22 | 638 |
| | | | | 27 | 614 |
| | | | | 3 Oct. | 632 |
| | | | | 24 | 628 |
| | | | | 25 | 632 |
| | | | | 28 | 630 |
| | | | | 29 | 626 |
| | | | | 5 Nov. | 626 |
| | | | | 8 | 644 |
| | | | | 16 | 634 |
| | | | | 17 | 631 |
| | | | | 23 | 629 |
| | | | | 30 | 618 |
| | | | | 10 Dec. | 629 |
| | | | | 11 | 622 |
| | | | | 22 | 601 |

Table 3E: Total Incident Solar Radiation for Cloudless Days
Mahé Island, Seychelles Islands (ly day$^{-1}$)

| 1963 | | 1964 | | | | | | 1965 | |
|---|---|---|---|---|---|---|---|---|---|
| 4 Nov. | 667 | 2 Jan. | 669 | 10 | 580 | 13 Aug. | 601 | 19 Jan. | 682 |
| 13 | 674 | 6 Feb. | 673 | 22 | 551 | 20 Sept. | 675 | 20 | 684 |
| 26 Dec. | 686 | 7 | 668 | 28 | 572 | 12 Oct. | 682 | 10 Mar. | 694 |
| | | 8 | 684 | 29 | 558 | 27 | 694 | 11 | 701 |
| | | 9 | 658 | 30 | 556 | 24 Nov. | 680 | 26 | 665 |
| | | 24 Mar. | 671 | 6 June | 536 | 25 | 682 | 2 Apr. | 675 |
| | | 2 Apr. | 637 | 11 | 561 | 5 Dec. | 675 | | |
| | | 4 | 595 | 2 July | 547 | 13 | 673 | | |
| | | 1 May | 609 | 25 | 597 | | | | |

Table 3F: Total Incident Solar Radiation
for Cloudless Days
Christmas Island (ly day$^{-1}$)

| 1963 | | 1964 | | 1965 | |
|---|---|---|---|---|---|
| 30 Aug. | 573 | 17 Jan. | 690 | 20 Jan. | 677 |
| 2 Sept. | 584 | 24 | 679 | 6 Feb. | 680 |
| 27 Oct. | 644 | 6 Feb. | 669 | 25 Mar. | 617 |
| 22 Nov. | 653 | 20 Mar. | 653 | 23 Apr. | 562 |
| 30 | 661 | 9 Apr. | 592 | 15 May | 521 |
| 25 Dec. | 693 | 8 May | 559 | 10 June | 445 |
| | | 4 June | 485 | 30 | 463 |
| | | 21 July | 497 | 21 July | 444 |
| | | 29 Sept. | 626 | 29 | 451 |
| | | 2 Oct. | 628 | 12 Aug. | 445 |
| | | 20 | 661 | 18 | 513 |
| | | 12 Nov. | 635 | 23 | 531 |
| | | 14 Dec. | 637 | 1 Sept. | 510 |
| | | | | 9 | 550 |
| | | | | 10 | 550 |
| | | | | 1 Oct. | 633 |
| | | | | 25 | 620 |
| | | | | 10 Nov. | 654 |

Table 3G: Total Incident Solar Radiation
for Cloudless Days
Plaisance, Mauritius Island (ly day$^{-1}$)

| 1963 | | 1964 | | 1965 | |
|---|---|---|---|---|---|
| 21 June | 381 | 15 Jan. | 705 | 6 Jan. | 767 |
| 24 | 380 | 4 Feb. | 682 | 26 | 757 |
| 21 July | 398 | 23 | 662 | 23 June | 368 |
| 25 | 429 | 1 Apr. | 563 | 24 | 361 |
| 10 Aug. | 451 | 3 | 532 | 26 | 376 |
| 14 | 459 | 9 | 497 | 11 July | 391 |
| 15 | 464 | 5 May | 466 | 3 Aug. | 428 |
| 19 Sept. | 529 | 15 | 478 | 15 | 459 |
| 27 | 599 | 20 June | 346 | 2 Oct. | 667 |
| 9 Oct. | 644 | 23 | 399 | 14 | 642 |
| 3 Dec. | 702 | 28 | 409 | 21 | 644 |
| 18 | 712 | 14 Aug. | 493 | 22 | 645 |
| 29 | 709 | 20 Sept. | 630 | 19 Nov. | 685 |
| | | 3 Oct. | 635 | 25 | 682 |
| | | 15 | 674 | 5 Dec. | 709 |
| | | 21 | 698 | 12 | 695 |
| | | 12 Nov. | 760 | | |
| | | 16 | 749 | | |
| | | 19 | 747 | | |
| | | 23 | 738 | | |
| | | 25 | 727 | | |
| | | 26 | 721 | | |
| | | 27 | 750 | | |
| | | 20 Dec. | 745 | | |
| | | 24 | 746 | | |
| | | 26 | 726 | | |

Table 3H: Total Incident Solar Radiation for Cloudless Days
Fort Dauphin, Madagascar (ly day$^{-1}$)

| 1963 | | | | 1964 | | | | 1965 | |
|---|---|---|---|---|---|---|---|---|---|
| 9 Apr. | 494 | 24 | 456 | 4 Jan. | 731 | 4 Aug. | 365 | 10 Jan. | 733 |
| 8 June | 340 | 25 | 474 | 6 | 742 | 5 | 395 | 13 | 749 |
| 9 | 352 | 26 | 468 | 7 | 767 | 10 | 408 | 13 Feb. | 684 |
| 13 | 336 | 27 | 473 | 8 | 747 | 16 | 449 | 14 | 666 |
| 14 | 334 | 31 | 466 | 20 | 745 | 21 | 417 | 15 | 678 |
| 15 | 339 | 1 Sept. | 471 | 25 | 746 | 24 | 412 | 18 | 654 |
| 16 | 319 | 4 | 509 | 2 Feb. | 693 | 28 | 477 | 3 Mar. | 608 |
| 22 | 331 | 5 | 504 | 17 | 653 | 3 Sept. | 461 | 7 | 582 |
| 23 | 338 | 6 | 502 | 18 | 660 | 11 | 506 | 28 | 563 |
| 24 | 308 | 7 | 511 | 11 Mar. | 574 | 17 | 524 | 29 | 519 |
| 28 | 331 | 9 | 489 | 20 | 592 | 18 | 516 | 26 May | 355 |
| 9 July | 315 | 10 | 477 | 28 | 547 | 25 | 562 | 27 | 327 |
| 10 | 321 | 11 | 445 | 16 Apr. | 473 | 28 | 574 | 28 | 334 |
| 11 | 325 | 13 | 482 | 27 | 443 | 14 Oct. | 618 | 29 | 335 |
| 13 | 332 | 17 | 509 | 28 | 439 | 15 | 598 | 30 | 342 |
| 15 | 332 | 18 | 528 | 6 May | 433 | 18 | 598 | 21 June | 336 |
| 19 | 356 | 25 | 524 | 7 | 388 | 20 | 604 | 24 | 337 |
| 20 | 356 | 4 Oct. | 564 | 8 | 397 | 25 | 623 | 9 Oct. | 589 |
| 25 | 354 | 6 | 566 | 11 | 424 | 9 Nov. | 649 | 16 | 634 |
| 26 | 353 | 20 | 635 | 12 | 393 | 10 | 658 | 18 | 622 |
| 30 | 410 | 22 | 658 | 13 | 391 | 14 | 677 | 27 | 649 |
| 31 | 384 | 12 Nov. | 718 | 15 | 369 | 15 | 667 | 28 | 628 |
| 1 Aug. | 391 | 16 | 662 | 22 | 356 | 17 | 644 | 29 | 648 |
| 6 | 394 | 28 | 752 | 12 June | 317 | 18 | 638 | 30 | 653 |
| 8 | 398 | 8 Dec. | 718 | 19 | 329 | 20 | 658 | 31 | 658 |
| 9 | 399 | 12 | 740 | 20 | 329 | 21 | 638 | 12 Nov. | 714 |
| 16 | 392 | 19 | 736 | 22 | 314 | 12 Dec. | 720 | 13 | 704 |
| 17 | 416 | 20 | 782 | 23 | 305 | 30 | 736 | 16 Dec. | 716 |
| 18 | 400 | 21 | 755 | 4 July | 325 | | | 23 | 720 |
| 21 | 440 | 22 | 745 | 20 | 361 | | | | |
| 22 | 454 | 29 | 773 | 25 | 362 | | | | |
| 23 | 466 | 30 | 771 | 29 | 360 | | | | |

Table 4A: Precipitable Water Data
       for Cloudless Days
       Port Blair, Andaman Islands (cm)

| 1963 | 1964 | 00 GMT | 12 GMT |
|------|------|--------|--------|
| 23 Dec. | | 2.83 | 2.72 |
| 24 | | 2.69 | 2.19 |
| | 9 Jan. | 3.41 | 3.79 |
| | 10 | 2.58 | |
| | 11 | | 2.98 |
| | 12 | 3.31 | 3.46 |
| | 25 | 2.77 | 3.69 |
| | 26 | 2.88 | 2.95 |
| | 28 | | 2.20 |
| | 29 | 2.24 | |
| | 31 | 2.12 | 2.85 |
| | 1 Feb. | 2.78 | 2.98 |
| | 6 | 2.43 | 2.83 |
| | 7 | | 3.27 |
| | 8 | 2.53 | |
| | 9 | 3.06 | |
| | 29 | 2.53 | 2.44 |
| | 2 Mar. | 3.23 | 3.26 |
| | 4 | 2.70 | |
| | 6 | 1.87 | |
| | 13 | 4.19 | 3.72 |
| | 15 | 4.09 | 4.04 |
| | 19 | 3.09 | 1.97 |
| | 20 | 2.75 | 3.40 |
| | 21 | 2.95 | 4.23 |
| | 23 | 3.79 | 3.58 |
| | 4 Apr. | 2.32 | |
| | 7 | 3.11 | |
| | 2 June | 4.02 | |
| | 8 Oct. | 4.44 | |
| | 10 Nov. | 3.95 | 3.98 |
| | 17 | 3.58 | 2.98 |
| | 5 Dec. | | 3.29 |
| | 11 | 3.55 | 3.27 |

Table 4B: Precipitable Water Data
       for Cloudless Days
       Trivandrum, India (cm)

| 1963 | 1964 | 00 GMT | 12 GMT |
|------|------|--------|--------|
| 13 Dec. | | 2.38 | |
| 14 | | 2.54 | 2.58 |
| 15 | | 3.03 | 3.02 |
| 18 | | 3.70 | 3.70 |
| 20 | | 3.75 | 3.43 |
| 28 | | 3.49 | 4.31 |
| 30 | | 3.29 | 3.87 |
| | 3 Jan. | 3.19 | 3.54 |
| | 4 | 3.21 | 4.48 |
| | 6 | 0.60 | 0.91 |
| | 8 | 3.29 | 4.50 |
| | 10 | 3.47 | 2.99 |
| | 13 | | 2.91 |
| | 19 | 3.49 | 3.07 |
| | 20 | 3.29 | 2.51 |
| | 21 | 3.88 | 3.69 |
| | 22 | 3.53 | 4.33 |
| | 4 Feb. | 2.88 | |
| | 5 | 2.80 | |
| | 6 | 2.23 | |
| | 7 | 2.52 | 3.77 |
| | 11 | 3.06 | 3.29 |
| | 12 | 2.39 | 2.81 |
| | 13 | 2.81 | 2.52 |
| | 14 | 2.57 | 3.73 |
| | 26 | 3.28 | 4.20 |
| | 9 Mar. | 3.26 | 3.08 |
| | 11 | 2.19 | 3.10 |
| | 13 | 2.67 | 3.89 |
| | 14 | 4.07 | 4.34 |
| | 19 | 2.95 | 3.59 |
| | 2 Apr. | 4.22 | |
| | 19 | 4.13 | |
| | 15 May | 3.75 | |
| | 16 | 3.13 | 5.10 |
| | 28 | 4.51 | |
| | 17 July | 4.38 | |
| | 17 Aug. | 3.97 | 4.93 |
| | 21 | | 3.43 |
| | 25 | 4.23 | |

Table 4C: Precipitable Water Data
for Cloudless Days
Mahé Island, Seychelles Islands (cm)

| 1963 | 1964 | 00 GMT |
|---|---|---|
| 4 Nov. | | 2.78 |
| 13 | | 4.34 |
| 26 Dec. | | 4.35 |
| | 2 Jan. | 3.61 |
| | 6 Feb. | 3.08 |
| | 7 | 3.04 |
| | 8 | 3.59 |
| | 9 | 4.02 |
| | 24 Mar. | 4.33 |
| | 2 Apr. | 3.95 |
| | 1 May | 2.79 |
| | 10 | 2.56 |
| | 22 | 3.59 |
| | 28 | 3.20 |
| | 29 | 3.84 |
| | 30 | 3.64 |
| | 6 June | 3.15 |
| | 11 | 2.51 |
| | 2 July | 2.89 |
| | 25 | 3.48 |
| | 13 Aug. | 2.33 |
| | 20 Sept. | 3.03 |
| | 12 Oct. | 3.23 |
| | 24 Nov. | 3.61 |
| | 25 | 3.35 |
| | 5 Dec. | 2.89 |
| | 13 | 3.45 |

Radiosonde data for Port Blair, Trivandrum, and Mahé Island (and other locations) were provided on punched cards by the International Meteorological Center, Bombay and were processed by digital computer to determine precipitable water by the method described in Appendix A. Data for other Indian Ocean area stations were processed also but could be used only on a comparative basis because they could not be associated directly with radiation recordings. Data were available for 1963 and 1964 for one (00 GMT) or two (00 and 12 GMT) soundings a day. Many soundings were missing or incomplete during the rainy season. Results of the precipitable water computations for Port Blair, Trivandrum, and Mahé Island are given in Table 4.

Precipitable water data were used with measured radiation data, known solar declination values, and the relationships given by Kimball (10) to compute fractional dust depletion for cloudless days. An average value for Port Blair and Trivandrum for December 1963 through March 1964 was about 0.05. Most computed values for Mahé Island, however, turned out to be slightly negative. The reason appeared to be the unusually large values of recorded solar radiation, a discussion of which is given below.

A latitudinal distribution of solar radiation for cloudless days throughout the year was determined from all curves labeled 2 in Figure 8 and is shown in Figure 9. Data for Mahé Island were excluded because of their questionably large values. The distribution was determined by (1) visually obtaining successive 10-day averages of curve 2 for each station throughout the year; (2) entering each 10-day average throughout the year opposite the latitude of each station on a composite graph for data from all stations; and (3) drawing isopleths of 10-day averages for each 25 ly day$^{-1}$. For the Northern Hemisphere, the isopleths were extrapolated from 13N, which is the latitude of Mangalore, to 25N. Monthly averages for each 5 degrees of latitude between 25N and 25S were then obtained from Figure 9 and are given in Table 5.

The curves in Figure 8 for the average measured solar radiation for cloudless days clearly show the effect of the variation of solar declination throughout the year for stations at different latitudes. Stations near the equator such as Mombasa had maxima of about 650 ly day$^{-1}$ in March and October, a pronounced minimum of about 500 ly day$^{-1}$ in June, and a secondary minimum of about 600 ly day$^{-1}$ in December. Fort Dauphin,

**FIGURE 8: Total incident solar radiation.**
*Curve 1:* Computed solar radiation incident on a horizontal surface at the outer limit of the earth's atmosphere for the latitude of the station. *Curve 2:* Estimated average of measured daily totals (indicated by crosses) at the surface for cloudless days. *Curve 3:* Computed solar radiation incident at the surface based on *Curve 1* and assumed values of precipitable water and fractional dust depletion as described below. *Curve 4:* Monthly averages of daily totals including cloudless days for 1964 as measured.

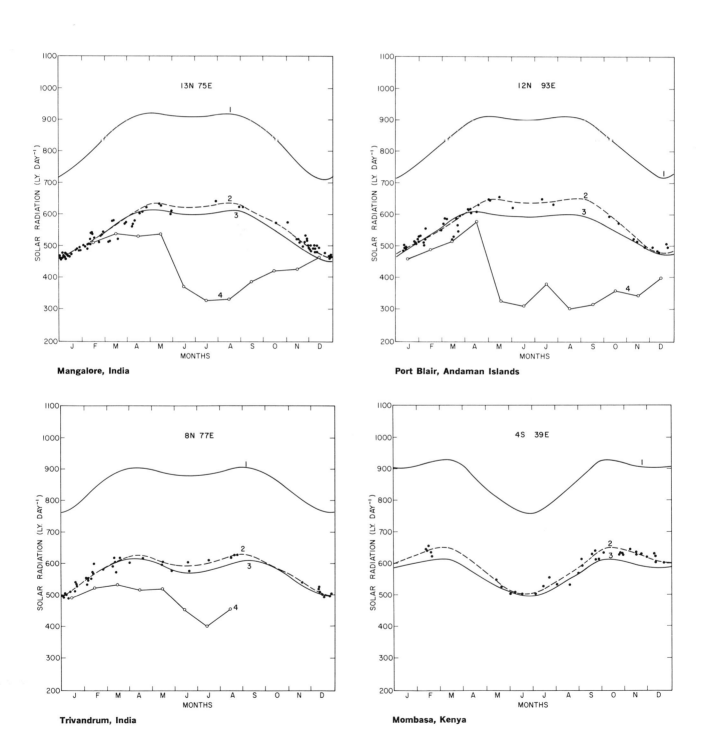

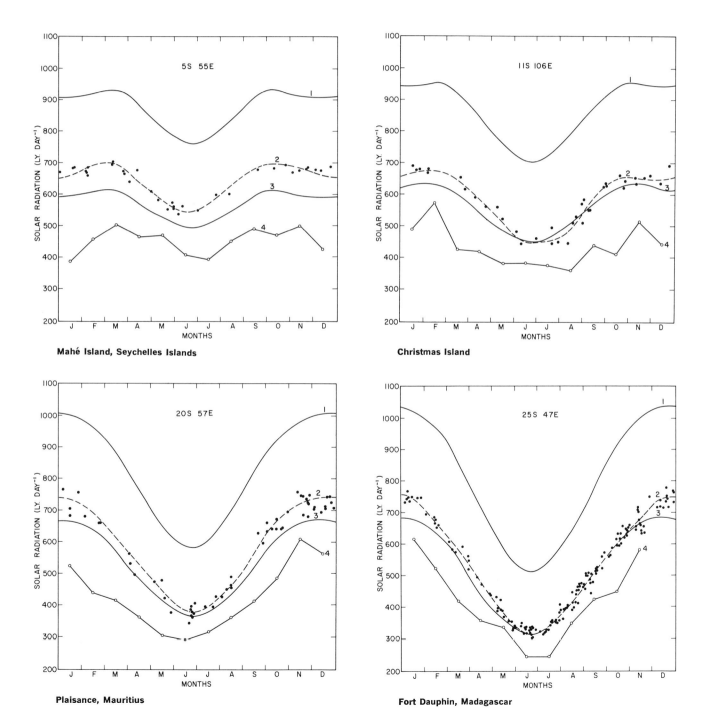

Mahé Island, Seychelles Islands

Christmas Island

Plaisance, Mauritius

Fort Dauphin, Madagascar

the southernmost station, had an average maximum of about 760 ly day$^{-1}$ in December and a minimum of about 320 ly day$^{-1}$ in June. Mangalore, the northernmost station, had maxima of about 640 ly day$^{-1}$ in May and August, a pronounced minimum of about 460 ly day$^{-1}$ late in December, and a secondary minimum of about 620 ly day$^{-1}$ in June.

It is evident from curves for Mombasa and Mahé Island that average daily totals for Mahé Island were about 50 ly day$^{-1}$ larger than those for Mombasa, even though both stations were at about the same latitude. Because Mahé is an island station located about 1750 km east of Mombasa on the coast of Africa, differences in both precipitable water and fractional dust depletion may be expected to account for some of the difference in solar radiation. No radiosonde data were available for Mombasa, but the average precipitable water for 1964 for Dar es Salaam, a coastal city about 300 km south of Mombasa, was 3.2 centimeters. The average value for Mahé Island was 3.4 centimeters. It seems unlikely, therefore, that the large difference in solar radiation would be the result of differences in precipitable water. Evidence to indicate that Mombasa had a larger average dust content, however, was given by Drummond and Vowinckel (11), who described the important influence of particulate matter in suspension in the atmosphere on the distribution of solar radiation for sections of southern Africa. They pointed out that extensive grass fires, especially in the large savannah lands, led to a significant reduction of incoming solar radiation during a nearly cloudless season over an area north of 30S.

Except for the likelihood that the dust content for Mombasa was greater, it was not possible to account completely for the much larger values of solar radiation for Mahé Island as compared to those obtained for Mombasa. Solar-radiation data for each station were compared for a few days with those obtained from identical sensors aboard research vessels when they were docked in these ports. Because they were not consistently larger or smaller, it could not be shown with certainty that equipment performance was responsible. Because the values for Mahé Island were disproportionately large for a latitude of 5S when included with data from the other stations in a latitudinal distribution and when compared to computed values shown in Figure 8, it was concluded that they were likely to be in error. Values for Mombasa, therefore, were used to represent solar radiation for cloudless days for a latitude of 5S.

Except for the results for Mahé Island and Mombasa, the longitudinal variation of daily totals for cloudless days was small. It can be noted from Figure 8, for example, that results for two other stations at nearly the same latitude — Port Blair in the Andaman Islands (12N, 93E) and Mangalore on the southwest coast of India (13N, 75E) — were very similar. Fragmentary data for Moroni in the Comores Islands (13S, 45E) also agreed with data for Christmas Island (11S, 106E). The above results led to the idea that, for the Indian Ocean, a latitudinal distribution of incident solar radiation for cloudless skies would have meaning independent of longitudinal variations. This idea was the basis for determining the latitudinal distribution of averaged daily totals throughout the year (Fig. 9).

Monthly averages of daily totals of incident solar radiation for cloudless days varied from about 5 per cent less for totals near 700 ly day$^{-1}$ to about 20 per cent less for totals near 350 ly day$^{-1}$ as compared to those tabulated by Budyko (3) for corresponding latitudes. The fact that Budyko's values were larger may be due to a combination of factors. One is the likelihood that the average precipitable water for cloudless days was actually greater for the present measurements. Budyko based his results largely on measurements made in locations where average values of precipitable water could easily have been less than for the locations used here. The latter were chosen, of course, so that the measurements would represent conditions for the maritime atmosphere of the Indian Ocean as closely as possible.

It is also likely that there was more attenuation owing to particulate matter for the present measurements than for Budyko's data. Mt. Agung, Bali (8S, 115E), erupted in March 1963 and discharged large amounts of volcanic material into the atmosphere. Anomalous attenuation of solar radiation late in 1963 and in 1964 was reported by Flowers and Viebrock (12) and by Budyko and Pivovarova (13), who hypothesized that volcanic dust from the Bali eruption was the responsible agent. The former authors reported reductions of solar radiation at normal incidence of 3 to 14 per cent of the normal intensity for Mauna Loa, Hawaii, and 23 to 78 per cent for the Antarctic. Similarly, Kondrat'yev et al. (14) attributed attenuation of solar radiation for his experiments in the Soviet Union in 1965 to the eruption of the volcano Taal on Luzon Island in September 1965. In each case, the amount of reduction depended on the solar zenith angle.

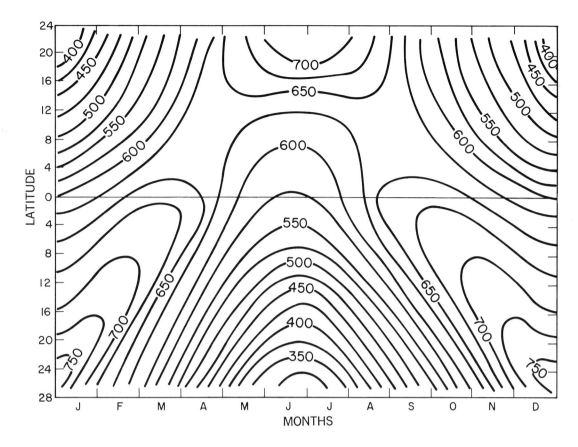

**FIGURE 9:** Latitudinal distribution of total incident solar radiation for cloudless days throughout the year. Isopleths of total incident solar radiation are drawn at intervals of 25 ly day⁻¹.

**Table 5:** Monthly Averages of Incident Solar Radiation for Cloudless Days for Each 5-Degree Latitude (ly day⁻¹)

| | Latitude | J | F | M | A | M | J | J | A | S | O | N | D |
|---|---|---|---|---|---|---|---|---|---|---|---|---|---|
| | 20–25 | 430 | 485 | 535 | 580 | 625 | 690 | 690 | 625 | 575 | 535 | 490 | 430 |
| | 15–20 | 455 | 500 | 550 | 590 | 635 | 670 | 655 | 630 | 585 | 550 | 500 | 450 |
| N | 10–15 | 480 | 525 | 570 | 620 | 640 | 635 | 640 | 650 | 610 | 575 | 525 | 485 |
| | 5–10 | 520 | 570 | 615 | 635 | 605 | 600 | 600 | 630 | 630 | 600 | 555 | 520 |
| | 0– 5 | 590 | 625 | 645 | 635 | 600 | 590 | 600 | 630 | 640 | 640 | 590 | 575 |
| | 0– 5 | 635 | 670 | 675 | 630 | 585 | 565 | 585 | 630 | 665 | 680 | 660 | 635 |
| | 5–10 | 660 | 690 | 680 | 610 | 555 | 530 | 550 | 600 | 655 | 685 | 690 | 665 |
| S | 10–15 | 695 | 710 | 650 | 580 | 510 | 450 | 490 | 540 | 625 | 680 | 710 | 690 |
| | 15–20 | 725 | 690 | 625 | 550 | 455 | 405 | 455 | 485 | 575 | 660 | 710 | 720 |
| | 20–25 | 735 | 660 | 580 | 500 | 415 | 350 | 415 | 440 | 520 | 610 | 690 | 740 |

A combination of the above factors, both causing greater attenuation of solar radiation, could reasonably account for the difference between the present results and those of Budyko. Because the first radiation recording station for the IIOE was placed into operation in April 1963, no data were obtained prior to the eruption on Bali. An analysis of measurements made before and after the eruption that might have quantified attenuation effects on account of the eruption was therefore not possible.

It can be noted by comparing curves 2 and 3 for each station that there were various levels of agreement between measured values and values obtained from Bolsenga's tables. Curve 2, representing measured values, is equal to or greater than curve 3 for all stations and times, except for Port Blair in March and for Christmas Island in July and August. The deviations are greatest during months of highest sun, and these correspond with rainy seasons in all cases, except for Mombasa. An expected increase in precipitable water for clear days during the rainy season, accounted for in the graphs for Trivandrum, Port Blair, and Mahé Island, but not in the others, tends to cause larger deviations in the annual pattern than those for constant precipitable water throughout the year. These factors suggest that seasonal variation in precipitable water is an unlikely cause for the observed differences between measured and computed values.

Seasonal variations in fractional dust depletion, not considered in preparing the graphs from Bolsenga's data, may be responsible for the deviations. Mani and Chacko (15) computed values of Angstrom's turbidity coefficient from measurements of solar radiation at normal incidence for Delhi and Poona and found that the largest turbidity coefficients (0.09 in June at Delhi and 0.04 in March at Poona) preceded the summer monsoon and that the smallest coefficients (0.02 in November at Delhi and 0.004 in October at Poona) followed it. A seasonal variation was not evident in the limited number of computations of dust content described above for Trivandrum and Port Blair.

## B. Incident Solar Radiation for Observed Cloudiness

The latitudinal distribution of daily totals of solar radiation for cloudless days given in Figure 9 and information on observed cloudiness were used to compute the distribution of solar radiation averaged for each month of 1963 and 1964. Cloudiness data were provided by the Interna-tional Meteorological Center on maps giving average cloudiness in tenths for each 5-degree latitude-longitude square for each month. Solar radiation data for cloudless days for each 5 degrees of latitude between 25S and 25N are given in Table 5 in terms of monthly averages of daily totals.

Several empirical relationships have been proposed for the dependence of average monthly solar radiation reaching a unit area at the earth's surface on average monthly cloudiness. A summary of most of the methods is given by Kondrat'-yev (16) and by Ta-k'ang, I-hsien, and Chien-sui (17). Of several relationships tested, the one that agreed best with our measurements was that proposed by Berlyand (18) and used by Budyko (3) for computations of solar radiation for each month on a world-wide basis. The relationship is

$$Q_s = Q_o(1-an-bn^2)$$

where

$Q_s$ = total solar radiation incident at the surface

$Q_o$ = total solar radiation incident at the surface, with a cloudless sky

$n$ = average cloudiness in tenths

and

$a, b$ = computed coefficients

Values of the coefficient $a$ were given by Berlyand (18) for different latitudes and varied between 0.35 and 0.4 for latitudes between 25S and 25N. He found that the coefficient $b$ was constant and equal to 0.38.

A comparison of results of the present measurements and the results computed by means of Berlyand's equation is shown in Figure 10. The ratio $Q_s/Q_o$, comprised of monthly averages of daily totals, is scaled on the abscissa and tenths of cloudiness on the ordinate. The points are data from different stations and are identified in the figure legend. The results of computations using Berlyand's equation for values of $a = 0.35$, 0.38, and 0.4 are shown as the three solid curves.

Figure 10 shows general agreement between measurements and computations, but a slightly greater decrease of $Q_s/Q_o$ for a given increase of cloudiness is indicated by the measurements. Some of the scatter of points and the departure from Berlyand's relationship mentioned above might have been due to inaccurate cloudiness estimates for each station. It was expected that estimates of cloudiness for specific stations from large-scale

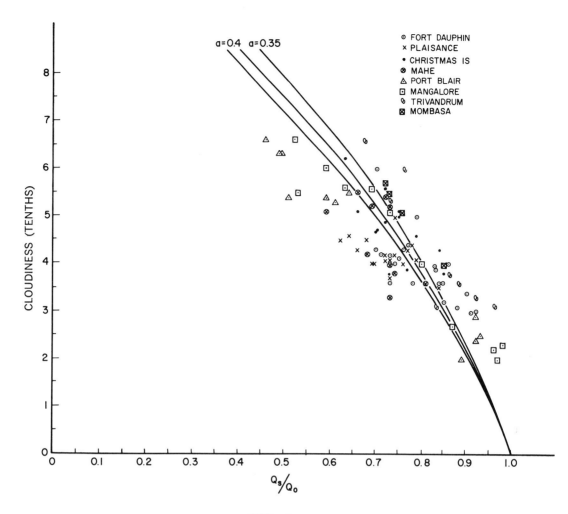

**FIGURE 10:** Measurements and computations of incident solar radiation for observed cloudiness. The abscissa is the ratio of solar radiation measured at the surface to that observed with a cloudless sky. The ordinate is tenths of cloudiness. The curves are based on Berlyand's relationship, and the symbols are measured values for the various stations identified in the legend.

cloudiness maps could differ significantly from actual cloudiness, even though all stations were located on coastlines. It is also possible that in some cases an insufficient number of cloud observations for a particular 5-degree latitude-longitude square gave a poor representation of the actual average monthly cloudiness for the square. Because exact effects of the above were not known and because a new relationship based on our measurements alone would be unreliable without actual cloud observations at each station, Berlyand's relationship was used to compute the distributions of solar radiation described below.

Monthly averages of daily totals of solar radiation were computed for each 5-degree latitude-longitude square between 25N and 25S and between about 40E and 110E. The results are shown in Figure 11 for each month of 1963 and 1964.

### C. Reflected Solar Radiation

To determine the distribution of reflected solar radiation in both space and time for cloudless areas, distributions of average diurnal albedo were computed. It has been established that the albedo of water depends primarily on water turbidity, roughness of the water surface, solar altitude, and

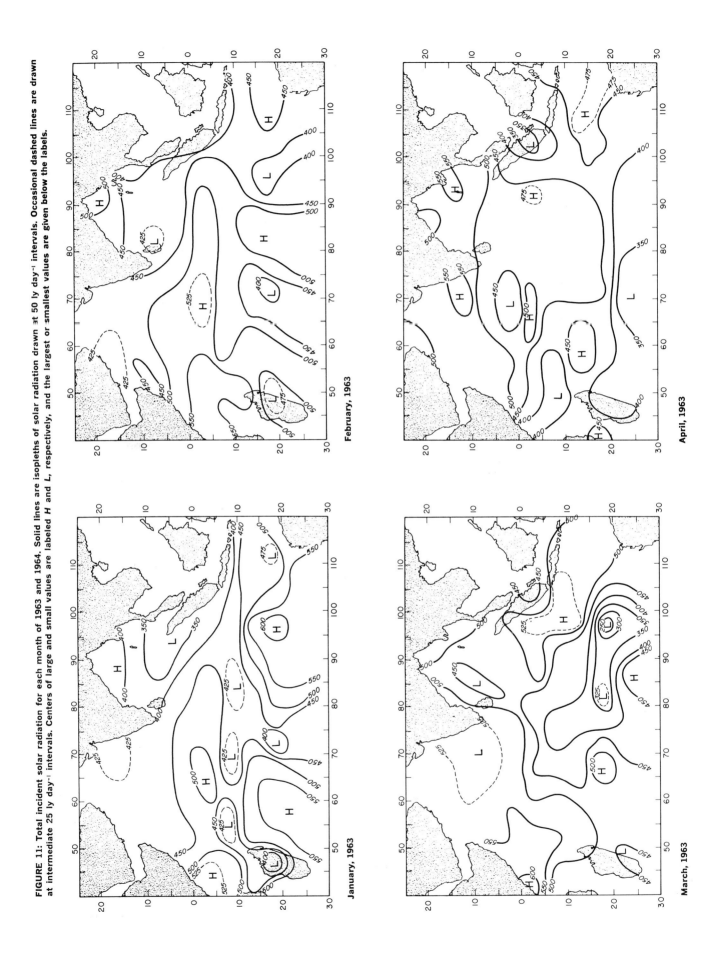

FIGURE 11: Total incident solar radiation for each month of 1963 and 1964. Solid lines are isopleths of solar radiation drawn at 50 ly day⁻¹ intervals. Occasional dashed lines are drawn at intermediate 25 ly day⁻¹ intervals. Centers of large and small values are labeled *H* and *L*, respectively, and the largest or smallest values are given below the labels.

January, 1963

February, 1963

March, 1963

April, 1963

June, 1963

August, 1963

May, 1963

July, 1963

**FIGURE 11:** *(Continued)*

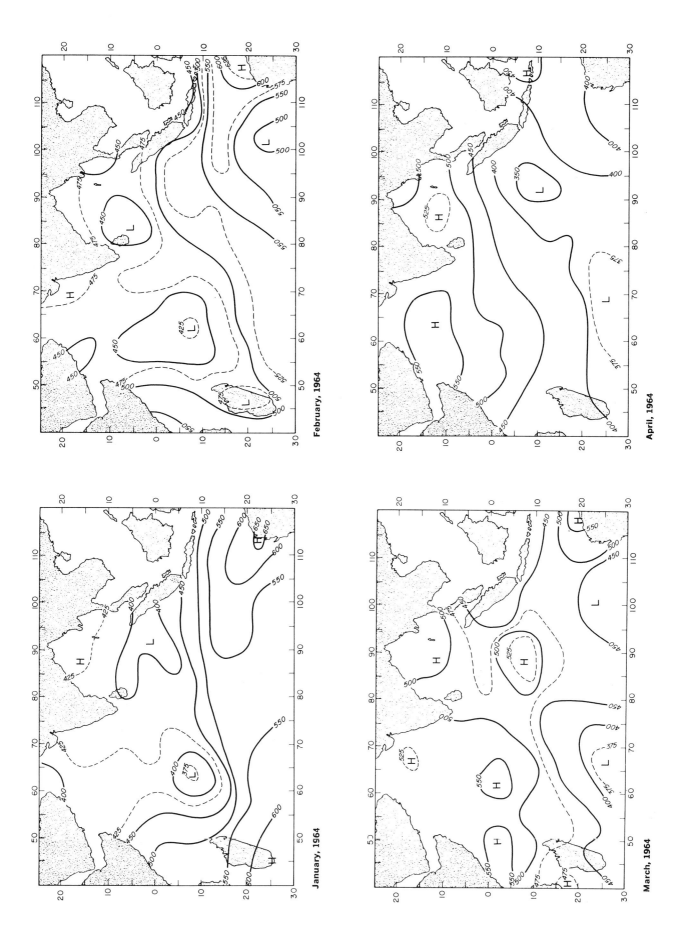

February, 1964

January, 1964

April, 1964

March, 1964

FIGURE 11: (Continued)

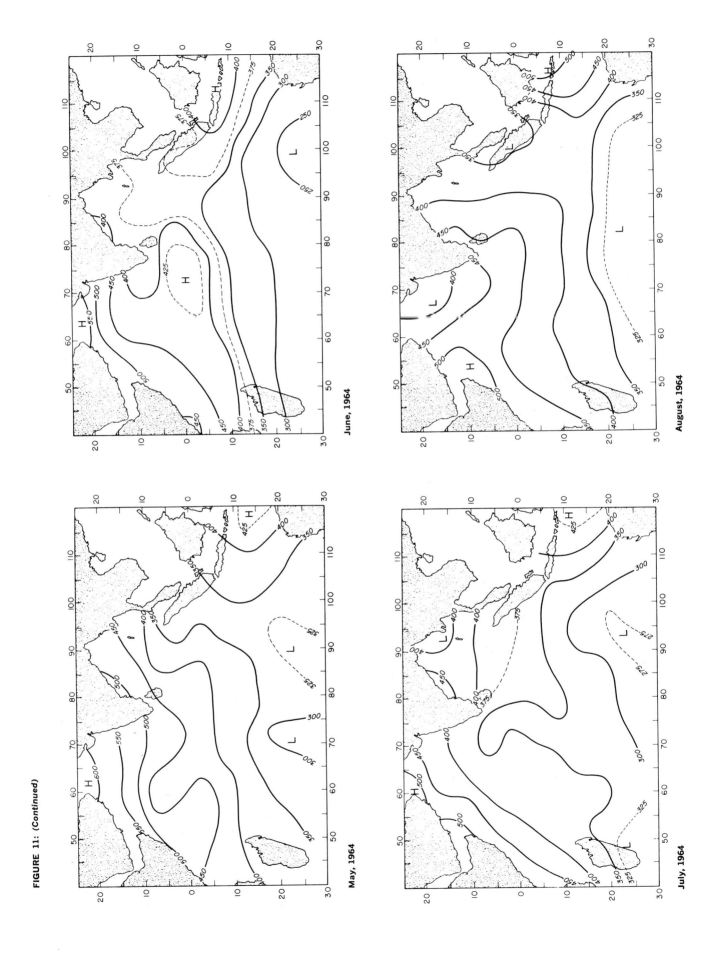

May, 1964

June, 1964

July, 1964

August, 1964

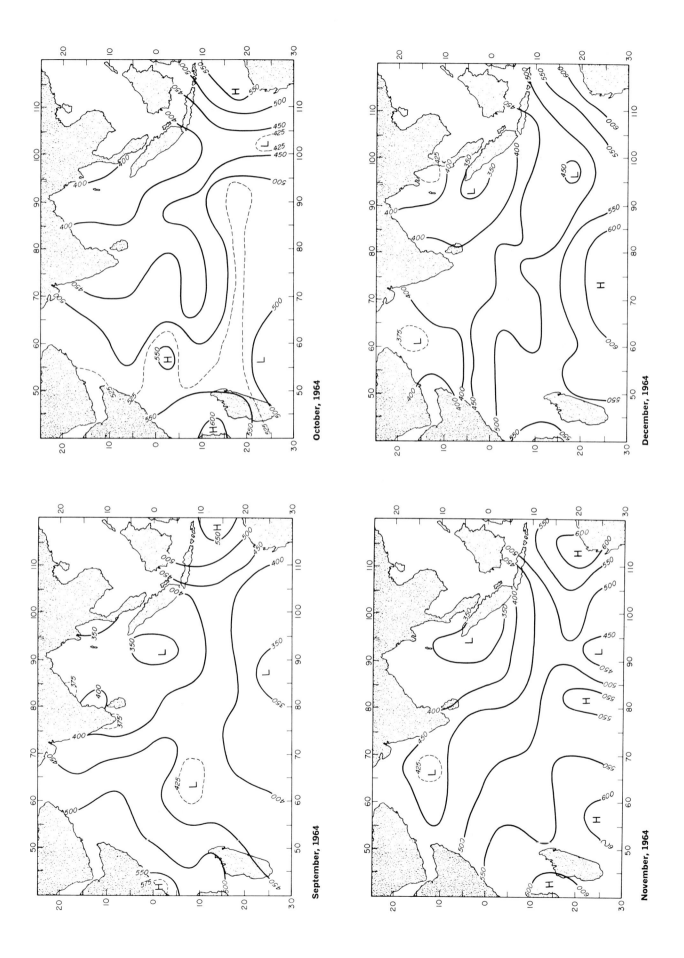

September, 1964

October, 1964

November, 1964

December, 1964

cloudiness. A discussion of albedo in relation to these variables is given by Kondrat'yev (16).

The effects of water turbidity were measured by Ter-Markaryants (19), who found that, for a solar altitude of 25 degrees, reflected solar radiation increased by a factor of 4 when its transparency decreased by a factor of 10. Actual values of reflected radiation, however, were small. They increased from 0.0015 ly min$^{-1}$ to 0.0065 ly min$^{-1}$ as transparency decreased from about 10 m to 1 meter. Estimates of effects of roughness vary considerably. Actual corrections for albedo for different solar altitudes and roughness were found by Burt (20) and Ter-Markaryants (19), but Degtyarev et al. (21) and Pivovarov et al. (22) did not find a significant effect of roughness up to a sea state of at least 3 (wave height of about 1.5 meters). Beard and Wiebelt (23) also concluded that waves have a very small effect on the magnitude of reflected solar radiation.

Because the effects of turbidity and roughness were uncertain and because reliable quantitative information on their distributions in space and time was unknown, these two factors were neglected in the computations of reflected solar radiation discussed below. The albedo for cloudless skies was computed only in relation to solar altitude.

There is general agreement among the results of various workers that albedo of water for a cloudless sky decreases as solar altitude increases. Observations of albedo for smooth nonturbid water have shown satisfactory agreement with computations based on the Fresnel equation except for very small solar altitudes. Results reported by Degtyarev et al. (21) and Pivovarov et al. (22), both of whom conducted albedo measurements on the Black Sea, are representative of the findings of others. The latter authors found excellent agreement between their data and the empirical expression

$$A = \frac{b}{\sin a + b},$$

where

$A$ = albedo,
$b$ = 0.04, an empirical constant, and
$a$ = solar altitude.

The above expression was used to calculate the average diurnal albedo for the geographical location of Fort Dauphin, Madagascar (25S, 47E); Port Blair, Andaman Islands (12N, 93E); and Plaisance, Mauritius Island (20S, 57E). Compu-

tations were made for at least two cloudless days for each station, one for which the solar declination was nearest to and one for which it was farthest from the latitude of each station.

Solar altitude was calculated by the expression

$$\sin a = \sin d \sin l + \cos d \cos l \cos h,$$

where

$a$ = solar altitude,
$d$ = solar declination,
$l$ = geographic latitude, and
$h$ = local hour angle.

Hourly albedo values were computed and multiplied by corresponding hourly averages of incident solar radiation as measured. The resulting hourly averages of reflected solar radiation were totaled for each day and divided by daily totals of incident solar radiation to obtain the average diurnal albedo.

The computations showed, as expected, that the average diurnal albedo was the smallest when the solar declination was nearest to the latitude of a station and that it was the largest when the declination was farthest from it. In other words, the albedo was least when the daily solar input was greatest; and, in fact, the relationship was found to be linear. These results are summarized in Table 6.

As solar radiation for cloudless days increased from approximately 300 to 700 ly day$^{-1}$ between the times of the winter and summer solstices for the Southern Hemisphere, as it did for Fort Dauphin, for example, the average diurnal albedo decreased from about 0.08 to 0.05. The corresponding increase of reflected solar radiation was about 12 ly day$^{-1}$. Stations closer to the equator had an even smaller change throughout the year. The averages given in Table 6 were about 7 ly day$^{-1}$ smaller than those found by Koberg (24), who analyzed the data of Anderson (25) and determined a relationship of reflected to measured solar radiation for both cloudless and cloudy skies.

Because of the relationship between albedo and incident daily totals, the latitudinal distribution of the latter transforms simply into a latitudinal distribution of reflected daily totals. Such a distribution is shown in Figure 12 as determined from the data in Figure 9 and Table 6.

### D. Net Radiative Exchange

In addition to maps of cloudiness, comparable maps of average monthly sea-surface temperatures were provided by the International Meteoro-

**Table 6: Reflected Solar Radiation for Cloudless Days (ly day⁻¹)**

| Incident Solar Radiation | Average Diurnal Albedo (in per cent) | Reflected Solar Radiation |
|---|---|---|
| 300 | 7.6 | 22.8 |
| 400 | 7.0 | 28.0 |
| 500 | 6.3 | 31.5 |
| 600 | 5.7 | 34.2 |
| 700 | 5.0 | 35.0 |

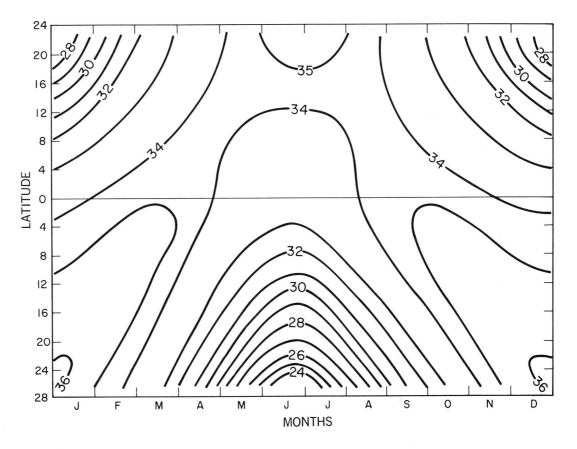

FIGURE 12: Latitudinal distribution of reflected solar radiation for cloudless days throughout the year. Isopleths of reflected solar radiation are drawn for every 1 ly day⁻¹.

logical Center in Bombay. The maps were the basis for computations of long-wave radiation emitted from the water surface. From this information, plus a knowledge of atmospheric radiation and net short-wave exchange for each 5-degree latitude-longitude square, distributions of net radiative exchange were computed for each month of 1963 and 1964. The procedure is described below.

The long-wave radiation emitted by the sea surface was computed by the Stefan-Boltzmann law, with monthly average sea-surface temperatures and an assumed water emissivity of unity. The smallest values, about 880 ly day$^{-1}$, were located approximately in the area 20S–25S and 70E–100E during July through October of 1963 and 1964. The largest average values, about 1005 ly day$^{-1}$, were located in the eastern half of the Arabian Sea in April and May of both years, just prior to the onset of the rainy season for that area.

The atmospheric component of the net long-wave exchange was estimated from measurements made at the island and coastal stations and from some measurements made aboard ships. Values of atmospheric radiation were obtained by subtracting values of solar radiation from corresponding values of total hemispherical radiation.

A study of atmospheric radiation in relation to limited information on cloudiness showed that daily totals of atmospheric radiation averaged about 720 ly (0.50 ly min$^{-1}$) and 835 ly (0.58 ly min$^{-1}$) for cloudless and cloudy conditions, respectively. Values between about 0.55 and 0.58 ly min$^{-1}$ (792 and 835 ly day$^{-1}$, respectively) were reported by Swinbank (26) based on shipboard measurements made between 10S and 22S in the Indian Ocean near Australia. Daily totals varying from about 740 to 880 ly for various amounts of cloudiness were reported by Hinzpeter (27) based on shipboard measurements made in the Red Sea.

There were insufficient atmospheric radiation data, unfortunately, to obtain reliable estimates of their seasonal and latitudinal variations. It appeared worthwhile, nonetheless, to consider the net radiation distribution that could be obtained by adopting a single value of atmospheric radiation — for all latitudes and months — and to interpret the results with the knowledge that a variability should be allowed for this term. After a review of all data on hand, it was decided that a value of 0.55 ly min$^{-1}$ (790 ly day$^{-1}$) for both daytime and nighttime conditions would be appropriate.

For computations of the net exchange of solar radiation, it was assumed that the average diurnal albedo of water was independent of cloudiness. The above assumption is not valid for small sun angles, as pointed out by Kondrat'yev (16); but his data show that for any month the mean diurnal albedo at 30 degrees latitude changes less than 1 per cent for an increase in low cloudiness from 0 to 0.6. The change was even less for latitudes closer to the equator. Cloudless albedos, therefore, were accepted as valid for all cloudiness. Monthly averages of daily totals of incident solar radiation for each 5-degree latitude-longitude square were multiplied by appropriate values of albedo to obtain net solar radiation for that square.

The above computations yielded for each 5-degree latitude-longitude square an average value, for each month, of net solar radiation and emitted long-wave radiation. Net solar radiation was added to atmospheric radiation taken to be 790 ly day$^{-1}$, and emitted long-wave radiation was subtracted from their sum to obtain the net radiative exchange. The distributions of net radiative exchange for each month of 1963 and 1964 are shown in Figure 13.

The distributions of areas of high and low net-radiative exchange for the same month in both 1963 and 1964 were similar, even though actual values for some months were quite different. The distributions for both Decembers and for both Junes, for example, were very similar. For December, low values were located over the Arabian Sea (less than 150 ly day$^{-1}$) and over the southern portion of the Bay of Bengal (less than 150 ly day$^{-1}$). High values were located east of the Seychelles Islands (greater than 250 ly day$^{-1}$) and across the Indian Ocean near 25S (greater than about 425 ly day$^{-1}$). A pronounced meridional gradient extended across the Indian Ocean between about 5S and 25S. The southward increase in net radiation is due to (1) an increase of net solar radiation from about 470 to 560 ly day$^{-1}$ and (2) a decrease of emitted long-wave radiation from about 980 to 930 ly day$^{-1}$.

For June, low values were located near the southwest coast of India (less than 190 ly day$^{-1}$), over the eastern portion of the Bay of Bengal (less than 150 ly day$^{-1}$), and over the southeast Indian Ocean in an area defined by about 5S to 25S and 85E to 105E (less than about 100 ly day$^{-1}$). High values were located over the northwest Arabian Sea (greater than 350 ly day$^{-1}$), south of India near the equator (greater than 210 ly day$^{-1}$), and near the east coast of India (greater than 200 ly day$^{-1}$). The gradient of net radiative exchange — instead of being nearly meridional as it was for December, with the highest values south of the equator — was directed northwest-southeast, with the highest values in the northwest Arabian Sea and the lowest values in an area about 1000 miles west-northwest of Australia. Net solar radiation decreased from about 540 to 230 ly day$^{-1}$, and emitted long-wave radiation decreased from about 980 to 940 ly day$^{-1}$ along the gradient.

For all months of 1963 and 1964, values of net radiative exchange were smaller than those found by Budyko (3) for the Indian Ocean. Although the distributions were similar, values obtained from the present measurements and computations were about 100 ly day$^{-1}$ smaller than Budyko's values for many areas. There was closer agreement with Mani et al. (4), both in the distributions of high and low areas and in their values, but most of our values were still somewhat smaller.

One of the main reasons for the difference between our values and those obtained by Budyko was that his values of incident solar radiation for cloudless days were larger, particularly for small values. Other factors contributing to the difference were: (1) our computations were based on data for 1963 and 1964, whereas Budyko's were based on climatological data; and (2) a value of 790 ly day$^{-1}$ for atmospheric radiation was accepted for the present work, whereas Budyko used climatological-atlas data on water-surface and air temperatures, water vapor, and cloudiness to compute net long-wave exchange. If atmospheric radiation actually increased from about 720 to 830 ly day$^{-1}$ from a cloudless to a cloudy area, as our data indicate, for example, the value of 790 ly day$^{-1}$ used in the present work could have caused the net radiative exchange to be about 70 ly day$^{-1}$ greater or 50 ly day$^{-1}$ less than the actual value for cloudless and cloudy areas, respectively.

The distributions of net radiative exchange for February and March 1964 can be compared to corresponding distributions of outgoing long-wave radiation (8 to 12 microns) measured above the atmosphere. The latter distributions are described by Winston (28), who analyzed measurements obtained by TIROS IV and TIROS VII. His maps of outgoing long-wave radiation for the same two months showed high values somewhat larger than 550 ly day$^{-1}$ extending across the Arabian Sea, India, and the Bay of Bengal and an elongated area of low values less than about 430 ly day$^{-1}$ extending from northeast of Madagascar to Indonesia. The distributions are very similar to corresponding distributions of net radiative exchange shown in Figure 13. Values of net radiative exchange were about 300 ly day$^{-1}$ for the former area and less than about 250 ly day$^{-1}$ for the latter.

The similarity of the above distributions was apparently determined to a large extent by similar dependencies of solar radiation, net radiative exchange, and outgoing long-wave radiation on cloudiness, which increased from an average of about 0.2 for the area of high values to about 0.5–0.6 for the area of low values. Net radiative exchange for a given month and latitude also varied inversely with cloudiness primarily because of the close dependence on incident solar radiation, which varied inversely with cloudiness as described above. A similar inverse correlation between cloudiness and outgoing long-wave radiation above the atmosphere, especially for low latitudes, was described by Adem (29), who found that, for latitudes of less than about 25 degrees, outgoing long-wave radiation decreased about 0.011 ly min$^{-1}$ for an increase of 0.1 in cloudiness.

## IX. SUMMARY AND CONCLUSIONS

With the aid of seven different national meteorological services, fourteen thermal radiation recording stations were established on islands and coasts of the Indian Ocean. Each station had an Eppley pyranometer (for total incident solar radiation), a modified Beckman and Whitley total hemispherical radiometer (for solar plus atmospheric radiation), and a special Minneapolis Honeywell twin recorder-integrator system. Considerable difficulty was experienced because of failure of electronic recorder components and intermittent and unreliable electrical power. Reliable data, however, have been obtained in sufficient amount to permit an analysis of the geographical distribution of average monthly radiation components of the heat exchange at the ocean surface for 1963 and 1964.

### A. Total Incident Solar Radiation

A latitudinal distribution of daily sums of incident solar radiation for cloudless days throughout the year was obtained from measurements at eight stations between 13N and 25S. The values were about 5 per cent less for totals near 700 ly day$^{-1}$ and about 20 per cent less for totals near 350 ly day$^{-1}$ than those obtained by Budyko (3). The differences may be due to the fact that an ab-

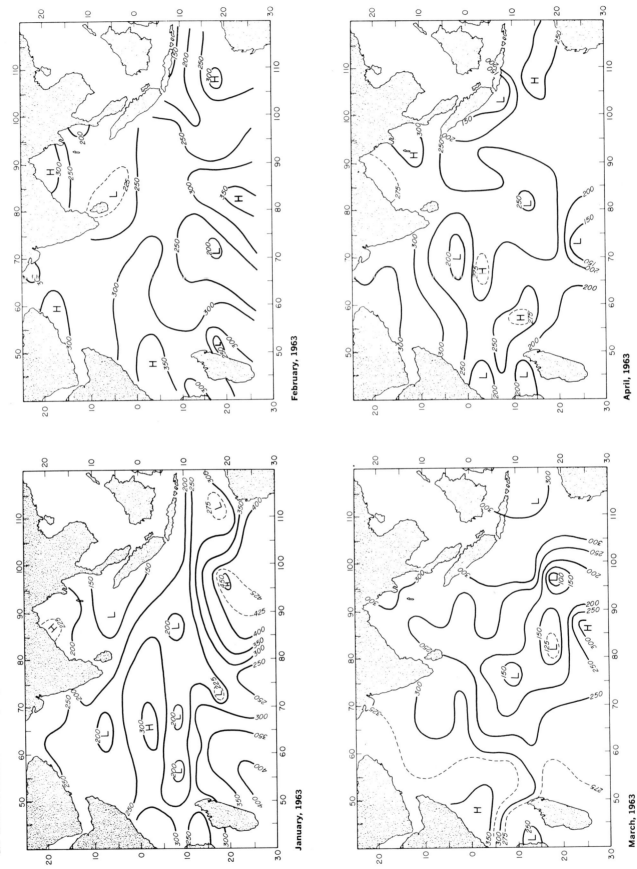

FIGURE 13: Net radiative exchange for each month of 1963 and 1964. Solid lines are isopleths of net radiation drawn at 50 ly day⁻¹ intervals. The dashed lines are drawn at intermediate 25 ly day⁻¹ intervals. Centers of large and small positive net radiative exchange are labeled with the letters H and L, respectively.

January, 1963

February, 1963

March, 1963

April, 1963

May, 1963

June, 1963

July, 1963

August, 1963

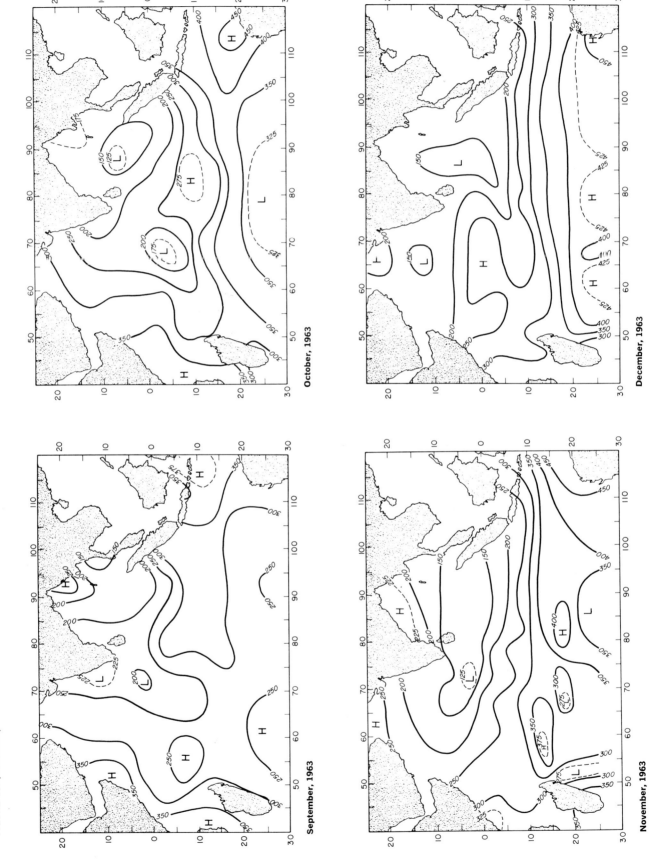

**FIGURE 13:** *(Continued)*

September, 1963

October, 1963

November, 1963

December, 1963

January, 1964

February, 1964

March, 1964

April, 1964

FIGURE 13: (Continued)

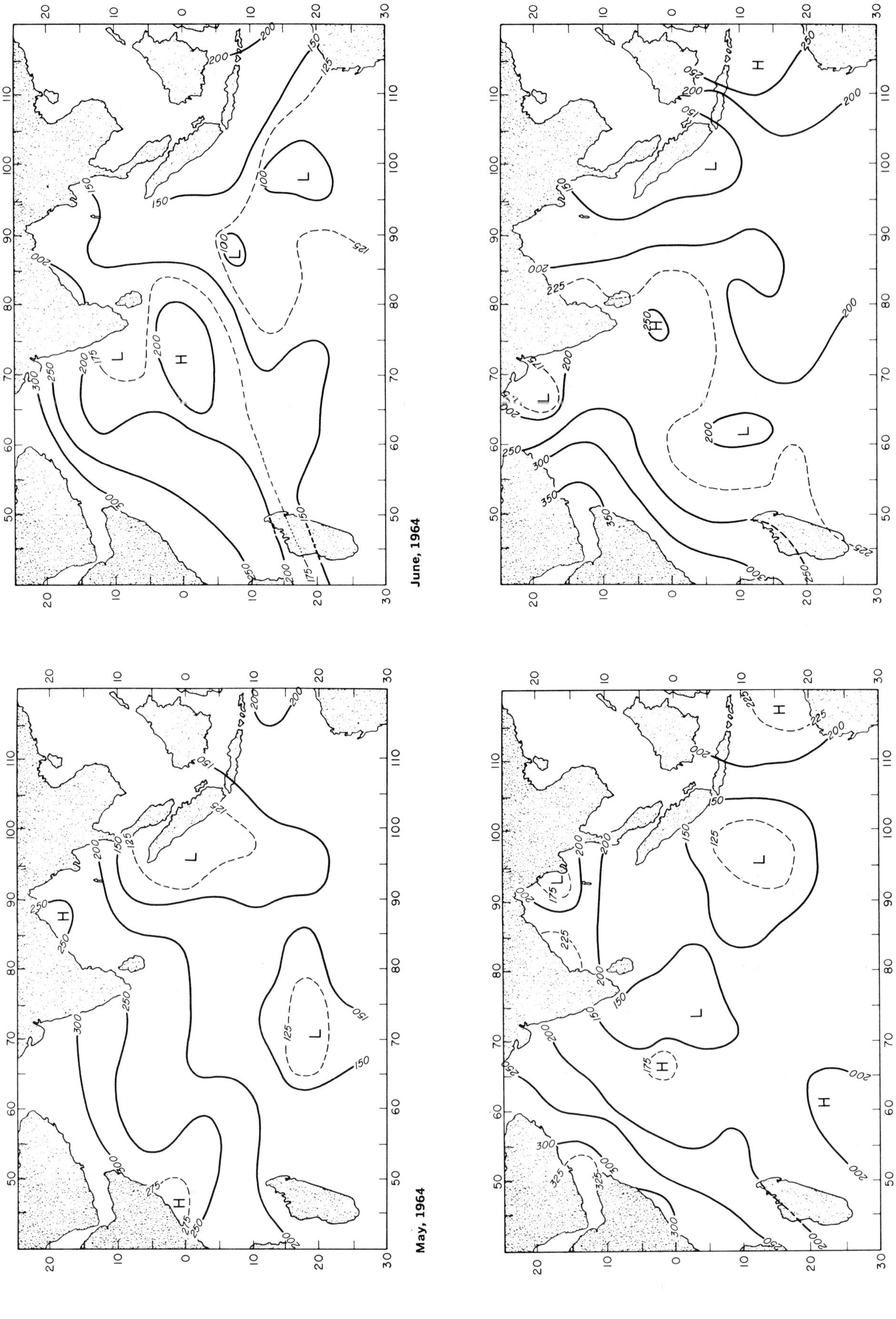

May, 1964

June, 1964

July, 1964

August, 1964

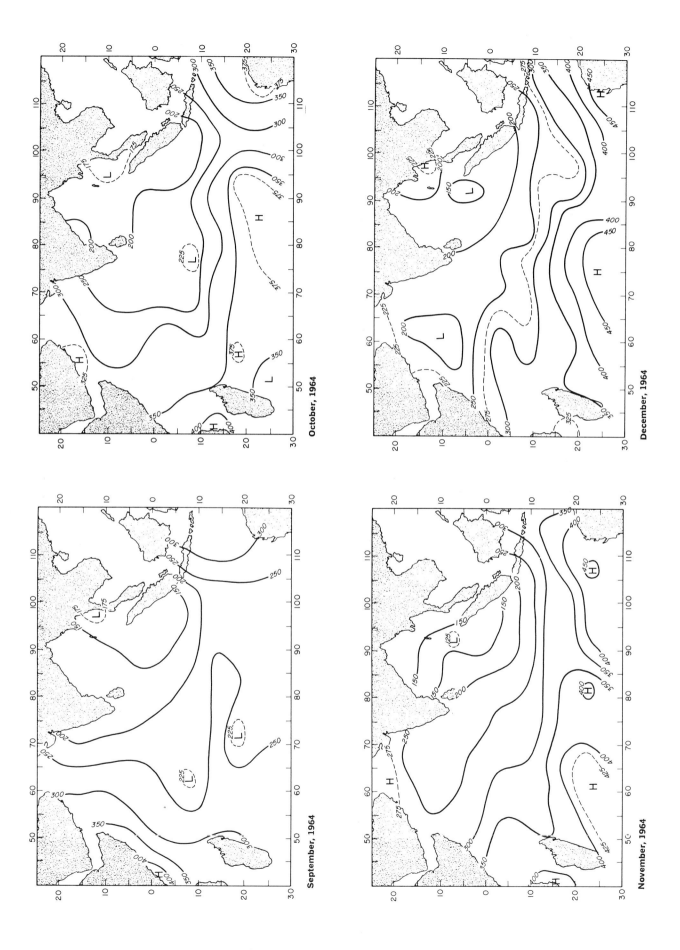

normally large amount of volcanic material was present in the atmosphere when our measurements were taken and also to the fact that our data may be more representative of moist tropical air masses than are those which Budyko used.

Distributions of average incident solar radiation for each month of 1963 and 1964 were computed from average cloudiness data for 5-degree latitude-longitude squares, values of solar radiation for cloudless days, and Berlyand's (18) relationship for estimating solar radiation from cloudiness. The results were similar to those obtained by Budyko (3) and by Mani et al. (4), all of whom used climatological cloudiness data. The values of the results obtained here, however, were smaller for the reasons given above.

### B. Reflected Solar Radiation

Values of reflected solar radiation were computed using values of incident solar radiation for cloudless days and an empirical relationship between albedo and solar altitude (Pivovarov et al., 22). A linear relationship between albedo and daily sums of incident solar radiation was obtained, which enabled daily sums of reflected solar radiation to be estimated from only daily sums of incident solar radiation. Values ranged from 23 to 35 ly day$^{-1}$ for values of incident solar radiation of 300 to 700 ly day$^{-1}$, respectively.

### C. Atmospheric Radiation

Daily sums of atmospheric radiation ranged from about 720 ly (0.50 ly min$^{-1}$) to 830 ly (0.58 ly min$^{-1}$) for cloudless and cloudy conditions, respectively. The most commonly recorded value, 0.55 ly min$^{-1}$ (790 ly day$^{-1}$), was accepted as a representative average for the Indian Ocean.

### D. Radiation Emitted by the Water Surface

The Stefan-Boltzmann relationship was used to compute daily sums of long-wave radiation emitted by the water surface corresponding to monthly average surface temperatures for each 5-degree latitude-longitude square. Values ranged from about 880 ly day$^{-1}$ in an area between about 20S and 25S and between 70E and 100E for July through October to about 1005 ly day$^{-1}$ in the eastern half of the Arabian Sea in April and May.

### E. Net Radiative Exchange

Distributions of net radiative exchange were computed for each month of 1963 and 1964 using information obtained from (A) through (D) above. Distributions obtained for the same months of both years were similar. For months for which corresponding distributions of outgoing long-wave radiation measured above the atmosphere were available, a similarity to the distributions of net radiative exchange was evident.

The results of the investigation have provided information on the seasonal and geographical distributions of various components of the radiative heat exchange between the atmosphere and the Indian Ocean and have, therefore, satisfied one of the goals of the International Indian Ocean Expedition.

# Appendix A: Computations of Precipitable Water

Precipitable water is defined (30) as the total atmospheric water vapor contained in a vertical column of unit cross-sectional area extending between two specified levels, commonly expressed in terms of the height to which that water substance would stand if completely condensed and collected in a vessel of the same unit cross section. It is given by

$$P.W. = \frac{1}{g} \int_{p_1}^{p_2} w\,dp, \qquad (1)$$

where

$P.W.$ = precipitable water (cm)
$g$ = acceleration of gravity (cm sec$^{-2}$)
$w$ = mixing ratio
$p$ = pressure (mb).

The mixing ratio is given to a good approximation by

$$w = \epsilon \frac{e}{p}, \qquad (2)$$

where $\epsilon = 0.622$, the ratio of the molecular weight of water vapor to the mean molecular weight of dry air, and $e$ = vapor pressure in millibars. According to an equation given by Tetens (31),

$$e = (6.11)(10)^{[7.5t_d/237.3+t_d]}, \qquad (3)$$

where $t_d$ = dew point in deg. C. The constants 7.5 and 237.3 are valid for water.

Combining the above equations gives

$$P.W. = \frac{\epsilon}{g\bar{p}} \int_{p_1}^{p_2} (6.11)(10)^{[7.5t_d/237.3+t_d]}dp. \qquad (4)$$

Radiosonde data were provided by the International Meteorological Center on punched cards.

Temperature and dew-point data were given for the surface, 1000 mb, and for every 50 mb above 1000 mb to the upper limit of each sounding. With $dp = 50$ mb, in equation (4), an average dew point for each 50 mb layer was obtained as described below. Figure 14 shows the pressure levels and corresponding dew-point levels. The slanted lines indicate the first three layers for which precipitable water was computed. The average dew point for the layer between the surface and 950 mb, $\bar{t}_d(1)$, was obtained as follows:

$$\bar{t}_d(1) = \frac{\left(\dfrac{t_d(1) + t_d(2)}{2}\right) + t_d(3)}{2}$$

and, for the layer between 900 and 950 mb,

$$\bar{t}_d(2) = \frac{t_d(3) + t_d(4)}{2}$$

where $t_d(1), t_d(2), t_d(3),$ and $t_d(4)$ are dew points for the surface, 1000 mb, 950 mb, and 900 mb levels, respectively.

By substituting the various constants into equation (4), it becomes for the first layer, for example,

$$P.W.(1) = \frac{(190)(10)^{[7.5\bar{t}_d(1)/237.3+\bar{t}_d(1)]}}{\bar{p}_1}.$$

To obtain the total amount of precipitable water for each sounding, values of precipitable water for all of the 50 mb layers were totaled. Only soundings having data extending to heights of at least 600 mb were used. A computer printout consisted of individual values for each of the 50 mb layers and of an over-all total for each sounding.

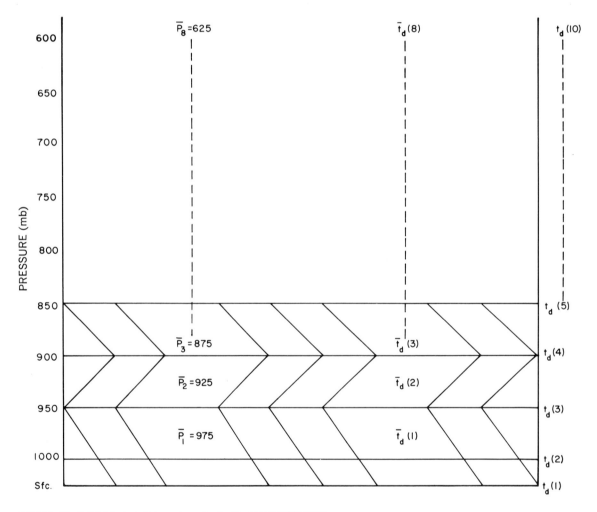

FIGURE 14: Pressure and temperature levels for precipitable-water computation.

Table 7: Solar Radiation, Karachi, West Pakistan (25N, 67E) (ly day$^{-1}$)

| 1963 | Jan. | Feb. | Mar. | Apr. | May | June | July | Aug. | Sept. | Oct. | Nov. | Dec. |
|------|------|------|------|------|-----|------|------|------|-------|------|------|------|
| 1  |  |  |     | 602 | 611 |     |     |  |  |  |  |  |
| 2  |  |  |     | 604 |     |     | 591 |  |  |  |  |  |
| 3  |  |  |     | 563 | 613 |     | 590 |  |  |  |  |  |
| 4  |  |  |     | 547 | 543 |     |     |  |  |  |  |  |
| 5  |  |  |     | 176 | 626 |     |     |  |  |  |  |  |
| 6  |  |  |     | 607 | 621 | 625 | 578 |  |  |  |  |  |
| 7  |  |  |     |     | 610 | 593 | 445 |  |  |  |  |  |
| 8  |  |  |     | 626 | 609 |     | 532 |  |  |  |  |  |
| 9  |  |  |     |     |     | 618 | 550 |  |  |  |  |  |
| 10 |  |  |     | 605 | 644 | 620 | 482 |  |  |  |  |  |
| 11 |  |  |     | 630 | 632 |     | 415 |  |  |  |  |  |
| 12 |  |  |     | 634 | 632 | 529 | 404 |  |  |  |  |  |
| 13 |  |  |     | 627 | 666 | 493 | 400 |  |  |  |  |  |
| 14 |  |  |     |     | 674 | 268 | 417 |  |  |  |  |  |
| 15 |  |  |     |     | 655 | 263 | 469 |  |  |  |  |  |
| 16 |  |  |     | 632 | 656 |     |     |  |  |  |  |  |
| 17 |  |  |     | 611 | 664 | 473 |     |  |  |  |  |  |
| 18 |  |  |     | 620 | 666 | 607 |     |  |  |  |  |  |
| 19 |  |  | 567 | 452 | 646 | 618 |     |  |  |  |  |  |
| 20 |  |  |     | 543 | 641 | 601 |     |  |  |  |  |  |
| 21 |  |  | 544 | 616 | 644 | 468 | 347 |  |  |  |  |  |
| 22 |  |  | 564 | 598 | 643 | 528 | 408 |  |  |  |  |  |
| 23 |  |  | 578 | 537 | 673 | 468 | 510 |  |  |  |  |  |
| 24 |  |  | 548 | 595 | 679 | 319 | 523 |  |  |  |  |  |
| 25 |  |  | 612 | 616 | 653 | 533 | 389 |  |  |  |  |  |
| 26 |  |  |     |     | 618 | 641 | 588 | 242 |  |  |  |  |
| 27 |  |  |     | 503 | 632 | 663 | 319 |  |  |  |  |  |
| 28 |  |  |     | 621 | 676 | 592 | 491 |  |  |  |  |  |
| 29 |  |  | 418 | 540 | 675 |     | 553 |  |  |  |  |  |
| 30 |  |  | 527 | 605 | 563 |     | 265 |  |  |  |  |  |
| 31 |  |  | 540 |     |     |     | 264 |  |  |  |  |  |

Table 8A: Solar Radiation, Mombasa, Kenya (4S, 39E) (ly day$^{-1}$)

| 1963 | Jan. | Feb. | Mar. | Apr. | May | June | July | Aug. | Sept. | Oct. | Nov. | Dec. |
|------|------|------|------|------|-----|------|------|------|-------|------|------|------|
| 1  |  |  |  |  |  | 310 | 309 | 298 | 355 | 560 |  |  |
| 2  |  |  |  |  |  | 226 | 226 | 335 | 492 | 518 |  |  |
| 3  |  |  |  |  |  |     | 305 | 244 | 441 | 395 |  |  |
| 4  |  |  |  |  |  | 242 | 372 | 367 | 334 | 507 |  |  |
| 5  |  |  |  |  |  | 231 | 450 | 347 | 541 | 291 |  |  |
| 6  |  |  |  |  |  | 541 | 348 | 453 | 429 | 508 |  |  |
| 7  |  |  |  |  |  | 627 | 323 | 332 | 513 | 474 |  |  |
| 8  |  |  |  |  |  | 568 | 365 | 486 | 496 | 621 |  |  |
| 9  |  |  |  |  |  | 594 | 409 | 491 | 483 | 582 |  |  |
| 10 |  |  |  |  |  | 375 | 162 | 514 | 368 |     |  |  |
| 11 |  |  |  |  |  | 337 | 446 | 373 | 484 |     |  |  |
| 12 |  |  |  |  |  | 581 | 464 | 403 | 424 |     |  |  |
| 13 |  |  |  |  |  |     | 485 | 323 | 383 |     |  |  |
| 14 |  |  |  |  |  |     | 416 | 418 | 321 |     |  |  |
| 15 |  |  |  |  |  | 385 | 467 | 531 | 478 |     |  |  |
| 16 |  |  |  |  |  | 276 | 449 | 527 | 425 |     |  |  |
| 17 |  |  |  |  |  | 441 | 355 | 556 | 479 |     |  |  |
| 18 |  |  |  |  |  | 206 | 430 | 339 | 393 |     |  |  |
| 19 |  |  |  |  |  | 136 | 421 | 392 | 440 |     |  |  |
| 20 |  |  |  |  |  | 392 | 349 | 364 | 566 |     |  |  |
| 21 |  |  |  |  |  | 272 | 422 | 399 | 547 |     |  |  |
| 22 |  |  |  |  |  | 413 | 471 | 287 | 506 |     |  |  |
| 23 |  |  |  |  |  | 395 | 485 | 275 | 602 |     |  |  |
| 24 |  |  |  |  |  | 455 | 389 | 194 | 544 |     |  |  |
| 25 |  |  |  |  |  |     | 412 | 503 | 357 |     |  |  |
| 26 |  |  |  |  |  | 216 | 363 | 387 | 577 |     |  |  |
| 27 |  |  |  |  |  |     | 512 | 521 | 584 |     |  |  |
| 28 |  |  |  |  |  |     | 184 | 258 | 466 | 477 |  |  |
| 29 |  |  |  |  |  | 224 | 350 | 367 | 271 | 389 |  |  |
| 30 |  |  |  |  |  | 460 | 344 | 425 | 577 |     |  |  |
| 31 |  |  |  |  |  |     | 257 | 362 |     |     |  |  |

Table 8B: Solar Radiation, Mombasa, Kenya (4S, 39E) (ly day$^{-1}$)

| 1964 | Jan. | Feb. | Mar. | Apr. | May | June | July | Aug. | Sept. | Oct. | Nov. | Dec. |
|---|---|---|---|---|---|---|---|---|---|---|---|---|
| 1 | | | | | | | | | | | | |
| 2 | | | | | | | | | | | | |
| 3 | | | | | | | | | | | | |
| 4 | | 568 | | | | | | | | | | |
| 5 | | 488 | | | | | | | | | | |
| 6 | | 539 | | | | | | | | | | |
| 7 | | 510 | | | | | | | | | | |
| 8 | | 552 | | | | | | | | | | |
| 9 | | 547 | | | | | | | | | | |
| 10 | | 506 | | | | | | | | | | |
| 11 | | 536 | | | | | | | | | | |
| 12 | | 604 | | | | | | | | | | |
| 13 | | 590 | | | | | | | | | | |
| 14 | | 560 | | | | | | | | | | |
| 15 | | 562 | | | | | | | | | | |
| 16 | | 555 | | | | | | | | | | |
| 17 | | 561 | | | | | | | | | | |
| 18 | | 617 | | | | | | | | | | |
| 19 | | 665 | | | | | | | | | | |
| 20 | | 590 | | | | | | | | | | |
| 21 | | | | | | | | | | | | |
| 22 | | | | | | | | | | | | |
| 23 | | | | | | | | | | | | |
| 24 | | | | | | | | | | | | |
| 25 | | | | | | | | | | | | |
| 26 | | | | | | | | | | | | |
| 27 | | | | | | | | | | | | |
| 28 | | 477 | | | | | | | | | | |
| 29 | | 389 | | | | | | | | | | |
| 30 | | | | | | | | | | | | |
| 31 | | | | | | | | | | | | |

Table 8C: Solar Radiation, Mombasa, Kenya (4S, 39E) (ly day$^{-1}$)

| 1965 | Jan. | Feb. | Mar. | Apr. | May | June | July | Aug. | Sept. | Oct. | Nov. | Dec. |
|---|---|---|---|---|---|---|---|---|---|---|---|---|
| 1 | | | | | | 466 | | | | | | |
| 2 | | | | | | 390 | 415 | 504 | 499 | 542 | 572 | 536 |
| 3 | | | | | | 241 | 438 | 263 | 508 | 595 | 176 | 432 |
| 4 | | | | | | 237 | 442 | 281 | 592 | 595 | 528 | 492 |
| 5 | | | | | 412 | 425 | 454 | 499 | 594 | 619 | 541 | 568 |
| 6 | | | | | 513 | 476 | 412 | 434 | 545 | 563 | 629 | 509 |
| 7 | | | | | 479 | 362 | 454 | 457 | 261 | 332 | 554 | 545 |
| 8 | | | | | 443 | 302 | 259 | 512 | 408 | 414 | 628 | 594 |
| 9 | | | | | 398 | 401 | 390 | 388 | 514 | 290 | 444 | 527 |
| 10 | | | | | 259 | 220 | 430 | 542 | 256 | 551 | 448 | 597 |
| 11 | | | | | 99 | 184 | 379 | 357 | 387 | 466 | 464 | 584 |
| 12 | | | | | 170 | 121 | 355 | 352 | 550 | 386 | 357 | 515 |
| 13 | | | | | 533 | 383 | 451 | 364 | 520 | 562 | 310 | 413 |
| 14 | | | | | 434 | 432 | 490 | 378 | 430 | 553 | 364 | 547 |
| 15 | | | | | 504 | 430 | 319 | 248 | 541 | 402 | 577 | 552 |
| 16 | | | | | 422 | 321 | 296 | 395 | 481 | 283 | 569 | 471 |
| 17 | | | | | 434 | 467 | 129 | 415 | 316 | 286 | 544 | 532 |
| 18 | | | | | 464 | 396 | 181 | 326 | 575 | 297 | 427 | 479 |
| 19 | | | | | 506 | 406 | 289 | 496 | 288 | 293 | 624 | 581 |
| 20 | | | | | 514 | 365 | 443 | 438 | 532 | 530 | 563 | 569 |
| 21 | | | | | 438 | 215 | 418 | 471 | 574 | 571 | 323 | 566 |
| 22 | | | | | 409 | 212 | 321 | 403 | 618 | 448 | | 598 |
| 23 | | | | | 382 | 382 | 403 | 462 | 290 | 383 | 600 | 561 |
| 24 | | | | | 407 | 410 | 368 | 350 | 541 | 593 | 572 | 568 |
| 25 | | | | | 490 | 439 | 467 | 533 | 537 | 570 | 439 | 497 |
| 26 | | | | | 232 | 407 | 404 | 409 | 576 | 597 | 507 | 484 |
| 27 | | | | | 344 | 445 | 256 | 516 | 568 | 604 | 445 | 388 |
| 28 | | | | | 481 | 435 | 214 | 355 | 308 | 617 | 522 | 560 |
| 29 | | | | | 472 | 429 | 527 | 470 | 568 | 590 | 536 | 596 |
| 30 | | | | | | 425 | 494 | 513 | 293 | 490 | 572 | 532 |
| 31 | | | | | | | 380 | 507 | | 387 | | 503 |

Table 8D: Total Hemispherical Radiation, Mombasa, Kenya (4S, 39E) (ly day$^{-1}$)

| 1963 | Jan. | Feb. | Mar. | Apr. | May | June | July | Aug. | Sept. | Oct. | Nov. | Dec. |
|---|---|---|---|---|---|---|---|---|---|---|---|---|
| 1 | | | | | | 1165 | | | | | | |
| 2 | | | | | 1071 | 1357 | | | | | | |
| 3 | | | | | | 1358 | | | | | | |
| 4 | | | | | | 1097 | | | | | | |
| 5 | | | | | | 1068 | | | | | | |
| 6 | | | | | | 1267 | | | | | | |
| 7 | | | | | | 1268 | | | | | | |
| 8 | | | | | | 1247 | | | | | | |
| 9 | | | | | | 1258 | | | | | | |
| 10 | | | | 1363 | | 1160 | | | | | | |
| 11 | | | | 1202 | | 1144 | | | | | | |
| 12 | | | | 1169 | | 1309 | | | | | | |
| 13 | | | | 1111 | | 1258 | | | | | | |
| 14 | | | | 1594 | | | | | | | | |
| 15 | | | | 1015 | | | | | | | | |
| 16 | | | | 1373 | | | | | | | | |
| 17 | | | | 1502 | | | | | | | | |
| 18 | | | | 1389 | | | | | | | | |
| 19 | | | | 1178 | | | | | | | | |
| 20 | | | | 1179 | | | | | | | | |
| 21 | | | | 1317 | | | | | | | | |
| 22 | | | | 903 | | | | | | | | |
| 23 | | | | 975 | | | | | | | | |
| 24 | | | | | | | | | | | | |
| 25 | | | | | | | | | | | | |
| 26 | | | | 1211 | 1070 | | | | | | | |
| 27 | | | | 1131 | 1230 | | | | | | | |
| 28 | | | | 1243 | 1153 | | | | | | | |
| 29 | | | | 1219 | 1283 | | | | | | | |
| 30 | | | | 1163 | 1232 | | | | | | | |
| 31 | | | | | 1210 | | | | | | | |

Table 8E: Total Hemispherical Radiation, Mombasa, Kenya (4S, 39E) (ly day$^{-1}$)

| 1964 | Jan. | Feb. | Mar. | Apr. | May | June | July | Aug. | Sept. | Oct. | Nov. | Dec. |
|---|---|---|---|---|---|---|---|---|---|---|---|---|
| 1 |  | 1435 |  | 1429 | 1382 | 1082 |  |  |  |  |  |  |
| 2 |  | 1378 | 1377 | 1495 | 1300 | 1195 |  | 1156 |  | 1487 | 1536 | 1397 |
| 3 |  | 1412 | 1455 | 1490 | 1363 | 1188 | 1157 | 1127 |  | 1430 | 1460 | 1256 |
| 4 |  | 1538 | 1518 | 1313 | 1348 | 1276 | 1086 | 1319 |  | 1328 | 1521 |  |
| 5 |  | 1427 | 1421 | 1065 | 1351 | 1297 | 1194 |  |  | 1444 | 1524 | 1192 |
| 6 |  | 1465 | 1414 | 1289 | 1347 | 1323 | 1173 | 1300 |  | 1372 | 1547 | 1242 |
| 7 |  | 1399 | 1447 | 1315 | 1279 | 1321 | 1249 | 1206 |  |  | 1528 | 1387 |
| 8 |  | 1454 | 1385 | 1529 | 1299 | 1036 | 1118 | 1196 |  |  | 1517 | 1418 |
| 9 |  | 1473 | 1402 | 1489 | 1325 | 1210 | 1194 | 976 |  | 1344 | 1548 | 1346 |
| 10 |  | 1450 | 1381 | 1547 | 1245 | 1259 | 1144 | 1352 |  | 1369 | 1519 | 1359 |
| 11 |  | 1486 | 1354 | 1335 | 1038 | 1200 | 1045 |  |  | 1199 | 1531 | 1429 |
| 12 |  | 1536 | 1461 | 1441 | 1355 | 1165 | 1011 |  |  |  | 1426 | 1367 |
| 13 |  | 1534 | 1512 | 1336 | 1050 | 1262 | 1144 |  |  | 1058 | 1477 | 1460 |
| 14 |  | 1511 | 1438 | 1408 | 1198 | 1093 | 1087 |  |  |  | 1427 | 1466 |
| 15 |  | 1496 | 1502 | 1344 | 1345 | 1171 | 1282 |  |  |  | 1535 | 1426 |
| 16 |  | 1465 | 1559 | 1024 | 1293 | 1011 | 1124 |  |  |  | 1540 | 1427 |
| 17 |  | 1459 | 1531 | 989 | 1312 | 1187 | 1231 |  |  | 1417 | 1517 | 1386 |
| 18 |  | 1494 | 1504 | 1354 | 1238 | 1057 | 1294 |  |  | 1143 | 1555 | 1429 |
| 19 |  | 1508 | 1492 | 1078 | 1328 | 1231 | 1252 |  |  | 1309 | 1450 | 1127 |
| 20 |  | 1511 | 1394 | 1169 | 1283 | 1229 | 1304 |  |  | 1309 | 1565 | 1303 |
| 21 |  | 1454 | 1220 | 1150 | 1184 | 1278 | 1193 |  |  | 1440 | 1583 | 1210 |
| 22 |  | 1312 | 1363 | 1136 | 1265 | 1303 | 1130 |  |  | 1368 | 1558 | 1114 |
| 23 |  | 1217 | 1431 | 1262 | 1093 | 1259 | 1141 |  |  | 1341 | 1447 | 1166 |
| 24 |  | 1532 | 1524 | 1183 | 1222 | 1243 | 1278 |  |  | 1461 | 1371 | 1012 |
| 25 |  | 1507 | 1359 | 1238 | 1169 | 1181 | 1208 |  |  | 1274 | 1423 | 1208 |
| 26 |  | 1514 | 1443 | 1301 | 1284 | 1181 | 1232 |  |  | 1361 | 1342 | 1374 |
| 27 |  | 1383 | 1418 | 1366 | 877 | 1124 | 1205 |  |  | 1394 | 1266 | 1243 |
| 28 | 1404 | 1581 | 1494 | 1396 | 1245 | 1175 | 1197 |  |  | 1405 | 1336 | 1387 |
| 29 | 1367 | 1320 | 1291 | 1304 | 1005 | 1217 | 1276 |  |  | 1419 | 1279 | 1387 |
| 30 | 1334 |  | 1544 | 1322 | 1179 | 1162 | 1286 |  |  |  |  | 1398 |
| 31 | 1417 |  | 1329 |  | 1256 |  |  |  |  | 1482 |  | 1366 |

Table 8F: Total Hemispherical Radiation, Mombasa, Kenya (4S, 39E) (ly day$^{-1}$)

| 1965 | Jan. | Feb. | Mar. | Apr. | May | June | July | Aug. | Sept. | Oct. | Nov. | Dec. |
|---|---|---|---|---|---|---|---|---|---|---|---|---|
| 1 | 1462 | 1477 | 1210 | | | | | | | | | |
| 2 | 1016 | 1345 | 1184 | 1129 | | | | | | | | |
| 3 | 1138 | 1397 | 1136 | 834 | | | | | | | | |
| 4 | 1283 | 1350 | 1139 | 1276 | | | | | | | | |
| 5 | 1318 | 1298 | 1135 | 1253 | | | | | | | | |
| 6 | 1326 | 1275 | 1034 | 1260 | | | | | | | | |
| 7 | 1358 | 1392 | 1026 | 1158 | | | | | | | | |
| 8 | 1232 | 1487 | 1082 | 1215 | | | | | | | | |
| 9 | 1307 | 1411 | 1124 | 1139 | | | | | | | | |
| 10 | 1425 | 1283 | | | | | | | | | | |
| 11 | | 1042 | | 986 | | | | | | | | |
| 12 | 1410 | | 1130 | 1150 | | | | | | | | |
| 13 | 1225 | | 1240 | 1161 | | | | | | | | |
| 14 | 1410 | | 1184 | 1159 | | | | | | | | |
| 15 | 1292 | | 1196 | | | | | | | | | |
| 16 | 1280 | | 1184 | | | | | | | | | |
| 17 | 1376 | | 1186 | | | | | | | | | |
| 18 | 1387 | | 1269 | | | | | | | | | |
| 19 | 1420 | 1052 | 1270 | | | | | | | | | |
| 20 | 1430 | 1114 | 1236 | | | | | | | | | |
| 21 | 1401 | 1016 | 1315 | | | | | | | | | |
| 22 | 1456 | 1114 | 1090 | | | | | | | | | |
| 23 | 1452 | 1136 | 1164 | | | | | | | | | |
| 24 | 1326 | 1245 | 1296 | | | | | | | | | |
| 25 | 1406 | 1204 | 1212 | | | | | | | | | |
| 26 | 1402 | 1259 | 1300 | | | | | | | | | |
| 27 | 1388 | 1073 | 1301 | | | | | | | | | |
| 28 | 1427 | 1199 | 1219 | | | | | | | | | |
| 29 | 1424 | | 1217 | | | | | | | | | |
| 30 | 1447 | | 1191 | | | | | | | | | |
| 31 | | | 1223 | | | | | | | | | |

Table 9A: Solar Radiation, Christmas Island (11S, 106E) (ly day$^{-1}$)

| 1963 | Jan. | Feb. | Mar. | Apr. | May | June | July | Aug | Sept | Oct | Nov | Dec |
|---|---|---|---|---|---|---|---|---|---|---|---|---|
| 1 | | | | | | | | | 467 | 590 | 606 | 542 |
| 2 | | | | | | | | | 565 | 589 | 414 | 566 |
| 3 | | | | | | | | | 563 | 454 | 550 | 415 |
| 4 | | | | | | | | | 501 | 464 | 538 | 560 |
| 5 | | | | | | | | | 329 | 585 | 498 | 476 |
| 6 | | | | | | | | | 244 | 512 | 584 | 528 |
| 7 | | | | | | | | | 254 | 530 | 528 | 578 |
| 8 | | | | | | | | | 549 | 538 | 410 | 635 |
| 9 | | | | | | | | | 553 | 602 | 432 | 609 |
| 10 | | | | | | | | | 363 | 558 | 379 | 448 |
| 11 | | | | | | | | | 444 | 629 | 367 | 472 |
| 12 | | | | | | | | | 550 | 514 | 350 | 550 |
| 13 | | | | | | | | | 414 | 585 | 387 | 599 |
| 14 | | | | | | | | | 503 | 475 | 544 | 577 |
| 15 | | | | | | | | | 533 | 358 | 548 | 591 |
| 16 | | | | | | | | | 577 | 349 | 500 | 351 |
| 17 | | | | | | | | | 400 | 479 | 299 | 341 |
| 18 | | | | | | | | | 332 | 388 | 324 | 486 |
| 19 | | | | | | | | | 319 | 432 | 416 | 448 |
| 20 | | | | | | | | | 488 | 529 | 570 | 435 |
| 21 | | | | | | | | | 340 | 368 | 585 | 310 |
| 22 | | | | | | | | 573 | 430 | 374 | 664 | 385 |
| 23 | | | | | | | | 558 | 525 | 233 | 560 | 299 |
| 24 | | | | | | | | 195 | 556 | 410 | 306 | 626 |
| 25 | | | | | | | | | 334 | 451 | 301 | 675 |
| 26 | | | | | | | | 541 | 520 | 497 | 386 | 629 |
| 27 | | | | | | | | 480 | 305 | 608 | | 660 |
| 28 | | | | | | | | 548 | 382 | 659 | 169 | 582 |
| 29 | | | | | | | | 365 | 360 | 611 | 535 | 635 |
| 30 | | | | | | | | 472 | 400 | 607 | | 488 |
| 31 | | | | | | | | 529 | | 403 | | 173 |

Table 9B: Solar Radiation, Christmas Island (11S, 106E) (ly day$^{-1}$)

| 1964 | Jan. | Feb. | Mar. | Apr. | May | June | July | Aug. | Sept. | Oct. | Nov. | Dec. |
|---|---|---|---|---|---|---|---|---|---|---|---|---|
| 1 | 245 | 670 | 611 | 483 | 476 | 361 | 445 | 466 | 385 | 594 | | 559 |
| 2 | 490 | 652 | 517 | 490 | 508 | 442 | 330 | 454 | 384 | 581 | | 560 |
| 3 | 349 | 550 | 355 | 494 | 270 | 393 | 146 | 333 | 405 | 202 | 437 | 412 |
| 4 | 281 | 536 | 458 | 500 | 313 | 466 | 291 | 464 | 172 | 123 | | 521 |
| 5 | 293 | 484 | 170 | 472 | 373 | 470 | 361 | 352 | 308 | 165 | 437 | 570 |
| 6 | 391 | 678 | 343 | 235 | 405 | 386 | 238 | 464 | 336 | 138 | 632 | 513 |
| 7 | 603 | 669 | 310 | 208 | 426 | 416 | 198 | 271 | 451 | 372 | 679 | 230 |
| 8 | 612 | 603 | 139 | 410 | 535 | 319 | 270 | 113 | 403 | 443 | 534 | 240 |
| 9 | 569 | 634 | 108 | 484 | 538 | 344 | 327 | 431 | 466 | 293 | 529 | 498 |
| 10 | 293 | 554 | 373 | 488 | 329 | 249 | 293 | 408 | 385 | 512 | 279 | 541 |
| 11 | 588 | 501 | 574 | 460 | 513 | 268 | 283 | 442 | 392 | 364 | 149 | 594 |
| 12 | 326 | 654 | 275 | 483 | 268 | 378 | 366 | 351 | 385 | 400 | 650 | 547 |
| 13 | 358 | 568 | 615 | 483 | 329 | 442 | 413 | 344 | 517 | 336 | 593 | 635 |
| 14 | 415 | 665 | 602 | 338 | 462 | 413 | 464 | 314 | 498 | 379 | 619 | 588 |
| 15 | 499 | 614 | 446 | 473 | 417 | 459 | 423 | 308 | 431 | 85 | 601 | 633 |
| 16 | | 678 | 494 | 418 | 286 | 376 | 346 | 291 | 370 | 250 | 650 | 503 |
| 17 | | 660 | 559 | 440 | 397 | 377 | 410 | 353 | 302 | 443 | 556 | 596 |
| 18 | 528 | 564 | 475 | 291 | 361 | 405 | 435 | 369 | 474 | 494 | 546 | 331 |
| 19 | 558 | 494 | 377 | 163 | 449 | 365 | 481 | 258 | 531 | 624 | 525 | 248 |
| 20 | 479 | 384 | 589 | 372 | 476 | 430 | 429 | 438 | 544 | 676 | 568 | 297 |
| 21 | 473 | | 495 | 449 | 363 | 401 | 444 | 381 | 466 | 673 | 285 | 267 |
| 22 | 451 | 371 | 479 | 412 | 320 | 374 | 460 | 331 | 462 | 612 | 571 | 308 |
| 23 | 444 | 528 | 535 | 503 | 101 | 382 | 398 | 335 | 547 | 308 | 420 | 416 |
| 24 | 649 | 493 | 289 | 473 | 357 | 467 | 266 | 257 | 461 | 398 | 454 | 550 |
| 25 | 632 | 492 | 317 | 359 | 320 | 182 | 447 | 264 | 522 | 569 | 550 | 311 |
| 26 | 668 | 481 | 232 | 370 | 467 | 294 | 408 | 433 | 630 | 522 | 615 | 195 |
| 27 | 651 | 488 | 485 | 467 | 385 | 375 | 437 | 470 | 406 | 628 | 516 | 130 |
| 28 | 689 | 651 | 320 | 460 | 302 | 448 | 439 | 429 | 546 | 543 | 540 | 403 |
| 29 | 533 | 674 | 469 | 466 | 386 | 391 | 484 | 323 | 602 | 278 | 580 | 455 |
| 30 | 424 | | 592 | 422 | 324 | 414 | 419 | 259 | | 304 | 470 | 384 |
| 31 | 626 | | 507 | | 312 | | 466 | 453 | | 235 | | |

Table 9C: Solar Radiation, Christmas Island (11S, 106E) (ly day$^{-1}$)

| 1965 | Jan. | Feb. | Mar. | Apr. | May | June | July | Aug | Sept | Oct | Nov. | Dec. |
|---|---|---|---|---|---|---|---|---|---|---|---|---|
| 1 | 617 | 310 | 299 | 285 | 442 | 436 | 452 | 470 | | 568 | 454 | 377 |
| 2 | 219 | 262 | 224 | 357 | 536 | 300 | 409 | 381 | 410 | 487 | 437 | 318 |
| 3 | 392 | 501 | 417 | 511 | 487 | 414 | 397 | 431 | 510 | 538 | 490 | 462 |
| 4 | 99 | 565 | 254 | 576 | 399 | 441 | 385 | 381 | 446 | 375 | 458 | 583 |
| 5 | 141 | 651 | 419 | 510 | 390 | 349 | 404 | 436 | 484 | 480 | 613 | 563 |
| 6 | 171 | 613 | 330 | 560 | 483 | 330 | 384 | 422 | 275 | 313 | 516 | 445 |
| 7 | 385 | 563 | 486 | 586 | 424 | 426 | 422 | 375 | 390 | 352 | 487 | 593 |
| 8 | 491 | 348 | 550 | 356 | 363 | 413 | 451 | 355 | 490 | 258 | 569 | 584 |
| 9 | 409 | 249 | 362 | 304 | 506 | 439 | 301 | 254 | 557 | 459 | 566 | 413 |
| 10 | 623 | 406 | 349 | 480 | 357 | 455 | 424 | 409 | 554 | 348 | 581 | 292 |
| 11 | 541 | 522 | 235 | 478 | 366 | 432 | 378 | 428 | 531 | 454 | 617 | 499 |
| 12 | 511 | 395 | 595 | | 442 | 444 | 333 | | 359 | 474 | 364 | 499 |
| 13 | 282 | 245 | 443 | 413 | 358 | 444 | 404 | 190 | 559 | 493 | 319 | 418 |
| 14 | 557 | 527 | 292 | 502 | 479 | 345 | 334 | 227 | 517 | 544 | 513 | 271 |
| 15 | 590 | 67 | 192 | 580 | 499 | 235 | 339 | 451 | 479 | 522 | 328 | 277 |
| 16 | 648 | 384 | 447 | 380 | 435 | 440 | 277 | 466 | 576 | 412 | 504 | 417 |
| 17 | 277 | 237 | 357 | 544 | 226 | 405 | 320 | 476 | 328 | 343 | 398 | 593 |
| 18 | 632 | | 565 | 524 | 86 | 457 | 306 | 436 | 482 | 522 | 450 | 337 |
| 19 | 350 | 421 | 238 | 503 | 181 | 457 | | 423 | | 534 | 586 | 257 |
| 20 | 562 | 504 | 497 | 443 | 214 | 412 | | 539 | | 387 | 481 | 276 |
| 21 | 424 | 576 | 521 | 496 | 414 | 440 | 426 | 549 | | 445 | 399 | 645 |
| 22 | 500 | 599 | 306 | 432 | 478 | 468 | 447 | 545 | | 289 | 351 | 589 |
| 23 | 572 | 643 | 281 | 570 | 413 | 442 | 395 | 528 | 529 | 427 | 321 | 417 |
| 24 | 527 | 593 | 599 | 507 | 262 | 300 | 403 | 526 | 336 | 327 | 302 | 296 |
| 25 | 641 | 530 | 595 | 430 | 495 | 306 | 324 | 397 | 357 | 548 | 401 | 569 |
| 26 | 500 | 366 | 543 | 501 | | 370 | 318 | 244 | 537 | 490 | 270 | 533 |
| 27 | 623 | 210 | 396 | 274 | 168 | 483 | 390 | | | 321 | 404 | 339 |
| 28 | 492 | 154 | 298 | 434 | 200 | 403 | 305 | | | 455 | 583 | 404 |
| 29 | 383 | | 282 | | 270 | 474 | 452 | | 448 | 416 | 510 | 333 |
| 30 | 331 | | 330 | | 271 | 464 | 449 | | 520 | 385 | 460 | 296 |
| 31 | 450 | | 521 | | 421 | | 400 | 314 | | 335 | | 256 |

Table 9D: Total Hemispherical Radiation, Christmas Island (11S, 106E) (ly day$^{-1}$)

| 1963 | Jan. | Feb. | Mar. | Apr. | May | June | July | Aug. | Sept. | Oct. | Nov. | Dec. |
|---|---|---|---|---|---|---|---|---|---|---|---|---|
| 1 | | | | | | | | | | | | 1251 |
| 2 | | | | | | | | | 1261 | 1307 | 1239 | 1283 |
| 3 | | | | | | | | | 1287 | 1274 | 1301 | 1220 |
| 4 | | | | | | | | | 1191 | 1286 | 1568 | 1351 |
| 5 | | | | | | | | | 1068 | 1370 | 1315 | 1250 |
| 6 | | | | | | | | | 936 | 1236 | 1363 | 1334 |
| 7 | | | | | | | | | 1068 | 1331 | | 1319 |
| 8 | | | | | | | | | 1221 | 1285 | 1158 | 1402 |
| 9 | | | | | | | | | 1248 | 1256 | 1201 | 1310 |
| 10 | | | | | | | | | 1140 | 1258 | 1184 | 1172 |
| 11 | | | | | | | | | 1166 | 1340 | 1292 | 1213 |
| 12 | | | | | | | | | 1242 | 1282 | 1149 | 1268 |
| 13 | | | | | | | | | 1236 | 1329 | | 1357 |
| 14 | | | | | | | | | 1254 | 1243 | 1349 | 1289 |
| 15 | | | | | | | | | 1293 | 1170 | | 1327 |
| 16 | | | | | | | | | 1251 | 1170 | | 1091 |
| 17 | | | | | | | | | 1077 | 1263 | 1150 | 1181 |
| 18 | | | | | | | | | 1071 | 1225 | | 1267 |
| 19 | | | | | | | | | 1113 | 1226 | | 1172 |
| 20 | | | | | | | | | 1209 | 1279 | 1320 | 1221 |
| 21 | | | | | | | | | 1097 | 1165 | 1333 | 1005 |
| 22 | | | | | | | | 1234 | 1179 | 1305 | 1363 | 1158 |
| 23 | | | | | | | | 1261 | 1182 | 1181 | | 1092 |
| 24 | | | | | | | | 928 | 1155 | 1267 | 1190 | 1302 |
| 25 | | | | | | | | | 1065 | 1264 | | 1332 |
| 26 | | | | | | | | 1266 | 1221 | 1277 | 1226 | 1440 |
| 27 | | | | | | | | 1244 | 1155 | 1348 | 1346 | 1362 |
| 28 | | | | | | | | 1253 | 1149 | 1364 | 1309 | 1347 |
| 29 | | | | | | | | 1231 | 1161 | 1342 | 1305 | 1392 |
| 30 | | | | | | | | 1242 | 1179 | 1356 | | 1248 |
| 31 | | | | | | | | 1240 | | 1219 | | |

Table 9E: Total Hemispherical Radiation, Christmas Island (11S, 106E) (ly day$^{-1}$)

| 1964 | Jan. | Feb. | Mar. | Apr. | May | June | July | Aug. | Sept. | Oct. | Nov. | Dec. |
|---|---|---|---|---|---|---|---|---|---|---|---|---|
| 1 | 1119 | | 1366 | | 1204 | 1185 | 1225 | 1189 | 1250 | | | 1359 |
| 2 | 1415 | 1246 | 1279 | | 1277 | 1197 | 1194 | 1181 | 1212 | | | | 1337 |
| 3 | 1133 | 1191 | 1106 | 1271 | 1095 | 1202 | 925 | 1071 | 1207 | | | | 1271 |
| 4 | 1147 | 1236 | 1251 | 1262 | 1178 | 1176 | 1048 | 1201 | | | | | 1514 |
| 5 | 1163 | 1057 | 926 | 1315 | 1197 | 1190 | 1122 | 1102 | 1207 | | | | 1438 |
| 6 | 1178 | 1356 | 1075 | 961 | 1207 | 1134 | 1025 | 1234 | 1203 | | | 1333 | 1443 |
| 7 | 1333 | 1324 | 1016 | 940 | 1230 | 1173 | 1000 | 1099 | 1258 | 1259 | 1428 | |
| 8 | 1296 | 1211 | 871 | 1173 | 1282 | 1098 | 1059 | 1006 | 1307 | 1146 | 1313 | |
| 9 | 1351 | 1289 | 848 | 1361 | 1227 | 1167 | 1124 | 1217 | 1326 | 1178 | 1332 | |
| 10 | 1061 | 1190 | 1151 | 1256 | 1033 | 1060 | 1084 | 1175 | 1259 | 1304 | 1122 | 1453 |
| 11 | 1324 | 1249 | 1340 | 1213 | 1244 | 1090 | 1107 | 1220 | 1290 | 1175 | 1256 | 1420 |
| 12 | | 1283 | 1016 | 1269 | 1080 | 1172 | 1249 | 1106 | 1224 | 1196 | 1389 | 1327 |
| 13 | 1115 | 1212 | 1339 | 1215 | 1091 | 1196 | 1237 | 1117 | 1357 | 1142 | 1297 | 1452 |
| 14 | 1193 | 1126 | 1336 | 1093 | 1242 | 1205 | 1279 | 1142 | 1352 | 1308 | 1298 | 1379 |
| 15 | 1254 | 1280 | 1169 | 1207 | 1233 | 1190 | 1220 | 1135 | 1321 | | 1290 | 1484 |
| 16 | | 1424 | 1249 | 1129 | 1060 | 1118 | 1168 | 1073 | 1269 | | 1350 | 1372 |
| 17 | | 1313 | 1285 | 1206 | 1195 | 1168 | 1180 | 1178 | 1276 | | 1223 | 1466 |
| 18 | 1289 | 1210 | 1194 | 1078 | 1132 | 1142 | 1201 | 1202 | 1361 | 1356 | 1212 | 1214 |
| 19 | 1239 | 1224 | 1105 | 960 | 1204 | 1077 | 1231 | 1103 | 1323 | 1387 | 1237 | |
| 20 | 1311 | 1062 | 1307 | 1138 | 1247 | 1128 | 1216 | 1201 | 1333 | 1391 | 1244 | |
| 21 | 1275 | | 1216 | 1214 | 1114 | 1202 | 1162 | 1164 | 1323 | 1372 | 953 | |
| 22 | 1247 | 1045 | 1221 | 1278 | 1150 | 1163 | 1189 | 1120 | 1267 | 1409 | 1295 | |
| 23 | 1346 | 1253 | 1280 | 1230 | 945 | 1170 | 1177 | 1116 | 1319 | 1292 | 1159 | 1258 |
| 24 | 1420 | 1187 | 1074 | 1147 | 1189 | 1172 | 1031 | 1045 | 1249 | 1384 | 1163 | |
| 25 | 1343 | 1157 | 1040 | 1121 | 1143 | 995 | 1221 | 1034 | 1277 | 1474 | 1314 | |
| 26 | 1403 | 1173 | 964 | 1080 | 1225 | 1096 | 1155 | 1192 | 1269 | 1417 | 1308 | |
| 27 | 1384 | 1189 | 1213 | 1240 | 1175 | 1232 | 1198 | 1166 | 1148 | 1387 | 1202 | |
| 28 | 1402 | 1323 | 971 | 1253 | 1051 | 1142 | 1160 | 1199 | 1306 | 1340 | 1230 | |
| 29 | 1293 | 1286 | 1244 | 1244 | 1168 | 1148 | 1257 | 1078 | 1357 | 1142 | 1266 | 1264 |
| 30 | 1210 | | 1328 | 1186 | 1130 | 1143 | 1210 | 1068 | | 1183 | 1151 | |
| 31 | 1397 | | 1249 | | 1091 | | | 1195 | | | | |

Table 9F: Total Hemispherical Radiation, Christmas Island (11S, 106E) (ly day$^{-1}$)

| 1965 | Jan. | Feb. | Mar. | Apr. | May | June | July | Aug. | Sept. | Oct. | Nov. | Dec. |
|---|---|---|---|---|---|---|---|---|---|---|---|---|
| 1 | | 1194 | | | | | | | | | | |
| 2 | 1060 | 1254 | | | | | | | | | | |
| 3 | | 1408 | | | | | | | | | | |
| 4 | | 1366 | | | | | | | | | | |
| 5 | | 1440 | | | | | | | | | | |
| 6 | 1081 | 1384 | | | | | | | | | | |
| 7 | 1145 | 1373 | | | | | | | | | | |
| 8 | 1382 | 1276 | | | | | | | | | | |
| 9 | 1234 | 1256 | | | | | | | | | | |
| 10 | 1550 | 1309 | | | | | | | | | | |
| 11 | 1354 | | | | | | | | | | | |
| 12 | 1298 | | | | | | | | | | | |
| 13 | 1119 | | | | | | | | | | | |
| 14 | 1368 | | | | | | | | | | | |
| 15 | 1385 | | | | | | | | | | | |
| 16 | 1406 | | | | | | | | | | | |
| 17 | 1077 | | | | | | | | | | | |
| 18 | 1417 | | | | | | | | | | | |
| 19 | 1070 | | | | | | | | | | | |
| 20 | 1317 | | | | | | | | | | | |
| 21 | 1170 | | | | | | | 1361 | | | | |
| 22 | 1304 | | | | | | | 1399 | | | | |
| 23 | 1287 | | | | | | | 1314 | | | | |
| 24 | 1361 | | | | | | | 1241 | | | | |
| 25 | 1400 | | | | | | | 1085 | | | | |
| 26 | 1306 | | | | | | | 1072 | | | | |
| 27 | 1444 | | | | | | | 1214 | | | | |
| 28 | 1279 | | | | | | | | | | | |
| 29 | 1162 | | | | | | | | | | | |
| 30 | 1171 | | | | | | | 1277 | | | | |
| 31 | 1340 | | | | | | | 1315 | | | | |

Table 10A: Solar Radiation, Fort Dauphin, Madagascar (25S, 47E) (ly day$^{-1}$)

| 1963 | Jan. | Feb. | Mar. | Apr. | May | June | July | Aug. | Sept. | Oct. | Nov. | Dec. |
|---|---|---|---|---|---|---|---|---|---|---|---|---|
| 1 | | | | | 323 | 252 | 246 | | 483 | | 674 | 457 |
| 2 | | | | | 443 | 269 | 178 | 390 | 199 | 567 | 225 | 436 |
| 3 | | | | | 412 | 238 | 148 | 195 | 476 | 524 | 456 | 356 |
| 4 | | | | | 354 | 291 | 158 | 310 | 518 | 564 | 464 | 363 |
| 5 | | | | | 254 | 280 | 191 | 280 | 503 | 467 | 562 | 563 |
| 6 | | | | | 332 | 160 | 232 | 399 | 513 | 566 | 511 | 586 |
| 7 | | | | | 421 | 215 | 258 | 325 | 515 | 533 | 196 | 586 |
| 8 | | | | | 421 | 311 | 244 | 398 | 484 | 196 | 729 | 628 |
| 9 | | | | | 242 | 295 | 310 | 400 | 502 | 527 | 527 | 680 |
| 10 | | | | | 321 | 328 | 321 | 337 | 483 | 237 | 441 | 216 |
| 11 | | | | | 396 | 359 | 326 | 290 | 448 | 230 | 587 | 563 |
| 12 | | | | | 405 | 331 | 137 | 292 | 495 | 548 | 707 | 740 |
| 13 | | | | | 399 | 337 | 332 | 316 | 488 | 496 | 503 | 518 |
| 14 | | | | | 397 | 334 | 203 | 411 | 475 | 558 | 314 | 409 |
| 15 | | | | | 400 | 344 | 332 | 353 | 205 | 550 | 532 | 463 |
| 16 | | | | | 367 | 320 | 222 | 383 | 468 | 336 | 746 | 382 |
| 17 | | | | | 178 | 265 | 344 | 416 | 512 | 236 | 728 | 379 |
| 18 | | | | | 394 | 108 | 344 | 400 | 526 | 337 | 733 | 656 |
| 19 | | | | | 209 | 189 | 355 | 248 | 356 | 636 | 421 | 736 |
| 20 | | | | 418 | 256 | 252 | 356 | 406 | 206 | 635 | 425 | 782 |
| 21 | | | | 298 | 385 | 332 | 88 | 440 | 526 | 596 | 428 | 755 |
| 22 | | | | 137 | 286 | 331 | 162 | 454 | 508 | 658 | 632 | 745 |
| 23 | | | | 130 | 379 | 338 | 352 | 466 | 457 | 610 | 461 | 364 |
| 24 | | | | 214 | 186 | 308 | 320 | 456 | 530 | 578 | 380 | 360 |
| 25 | | | | 356 | 318 | 200 | 354 | 474 | 525 | 376 | 632 | 319 |
| 26 | | | | 375 | 370 | 232 | 353 | 468 | 483 | 336 | 559 | 373 |
| 27 | | | | 134 | 398 | 291 | 121 | 473 | 284 | 350 | 709 | 183 |
| 28 | | | | 365 | 367 | 332 | 252 | 460 | 548 | 322 | 748 | 773 |
| 29 | | | | 247 | 376 | 260 | 322 | 362 | 559 | 504 | 312 | 771 |
| 30 | | | | 397 | 350 | 166 | 410 | 438 | 524 | 310 | 302 | 638 |
| 31 | | | | | 352 | | 396 | 465 | | 513 | | 560 |

Table 10B: Solar Radiation, Fort Dauphin, Madagascar (25S, 47E) (ly day$^{-1}$)

| 1964 | Jan. | Feb. | Mar. | Apr. | May | June | July | Aug. | Sept. | Oct. | Nov. | Dec. |
|---|---|---|---|---|---|---|---|---|---|---|---|---|
| 1 | 632 | 698 | | 329 | | | | | | | 654 | |
| 2 | 749 | 693 | 64 | 210 | 316 | 322 | 306 | 223 | | 479 | 308 | |
| 3 | 766 | 668 | 174 | 454 | 191 | 296 | 195 | 361 | 461 | 122 | 322 | |
| 4 | 740 | 485 | 471 | 482 | 382 | 66 | 325 | 374 | | 182 | 428 | |
| 5 | 577 | 274 | 621 | 218 | 195 | 153 | 298 | 395 | | 101 | 450 | |
| 6 | 742 | 492 | 628 | 409 | 433 | 191 | 263 | 307 | 406 | 154 | 495 | |
| 7 | 767 | 625 | 343 | 518 | 400 | 335 | 289 | 185 | 175 | 294 | 572 | |
| 8 | 747 | 283 | 575 | 527 | 397 | 306 | 197 | 251 | 211 | 403 | 634 | |
| 9 | 680 | 583 | 307 | 491 | 346 | 201 | 190 | 251 | 420 | 584 | 667 | |
| 10 | 537 | 578 | 421 | 454 | 394 | 311 | 238 | 408 | 502 | 349 | 657 | |
| 11 | 323 | 276 | 317 | 503 | 424 | 244 | 250 | 398 | 506 | 566 | 653 | |
| 12 | 164 | 524 | 256 | 391 | 393 | 317 | 189 | 391 | 413 | 321 | 644 | |
| 13 | 563 | 605 | 438 | 180 | 391 | 291 | 245 | 398 | 275 | 299 | 641 | |
| 14 | 575 | 565 | 343 | 178 | 335 | 107 | 328 | 164 | 439 | 618 | 677 | |
| 15 | 761 | 665 | 451 | 394 | 369 | 114 | 226 | 338 | 541 | 582 | 667 | |
| 16 | 619 | 676 | 613 | 472 | 391 | 145 | 189 | 449 | 519 | 421 | 674 | |
| 17 | 733 | 653 | 448 | 451 | 283 | 206 | 160 | 401 | 524 | 538 | 644 | |
| 18 | 618 | 663 | 590 | 189 | 430 | 262 | 311 | 426 | 580 | 598 | 638 | |
| 19 | 637 | 668 | 493 | 430 | 322 | 329 | 296 | 301 | 520 | 559 | 617 | |
| 20 | 745 | 217 | 592 | 458 | 337 | 329 | 361 | 347 | 541 | 604 | 658 | |
| 21 | 515 | 224 | 148 | 320 | 363 | 314 | 231 | 417 | 191 | 580 | 638 | |
| 22 | 277 | 493 | 551 | 175 | 356 | 314 | 231 | 352 | 467 | 529 | 588 | |
| 23 | 693 | 270 | 260 | 172 | 370 | 305 | 314 | 302 | 366 | 520 | | |
| 24 | 711 | 473 | 335 | 220 | 300 | 316 | 311 | 412 | 259 | 595 | | |
| 25 | 746 | 627 | 473 | 322 | 340 | 192 | 362 | 167 | 562 | 623 | | |
| 26 | 707 | 636 | 400 | 445 | 350 | 217 | 34 | 372 | 448 | 626 | | |
| 27 | 625 | 594 | 478 | 443 | 254 | 236 | 33 | 364 | 339 | 613 | | |
| 28 | 525 | 491 | 547 | 440 | 166 | 298 | 329 | 477 | 574 | 472 | | |
| 29 | 509 | 478 | 436 | 245 | 274 | 134 | 360 | 445 | 443 | 423 | | |
| 30 | 429 | | 487 | 253 | 298 | 194 | 212 | 394 | | 331 | | |
| 31 | 653 | | 289 | | 340 | | 27 | 460 | | | | |

Table 10C: Solar Radiation, Fort Dauphin, Madagascar (25S, 47E) (ly day$^{-1}$)

| 1903 | Jan. | Feb. | Mar. | Apr. | May | June | July | Aug. | Sept. | Oct. | Nov. | Dec. |
|------|------|------|------|------|-----|------|------|------|-------|------|------|------|
| 1 | | | | | | | | | | | | |
| 2 | | | | | | 415 | | | | | 573 | 412 |
| 3 | | | | | | 340 | | | | 485 | 618 | 518 |
| 4 | | | | | | 353 | | | | | 482 | 499 |
| 5 | | | | | | 385 | | | | 488 | 638 | 610 |
| 6 | | | | | | 343 | | | | 539 | 646 | 711 |
| 7 | | | | | | 431 | | | | 470 | 526 | 651 |
| 8 | | | | | | 275 | | | | 257 | 532 | 580 |
| 9 | | | | | | 158 | | | | 589 | 658 | 683 |
| 10 | | | | | | 325 | | | | 574 | 501 | 478 |
| 11 | | | | | | | | | | 604 | 670 | 347 |
| 12 | | | | | | 210 | | | | 532 | 714 | 563 |
| 13 | | | | | | 376 | | | | 69 | 704 | 709 |
| 14 | | | | | | 277 | | | | 162 | 611 | 742 |
| 15 | | | | | | 304 | | | | | 131 | 688 |
| 16 | | | | | | 345 | | | | 623 | 152 | 719 |
| 17 | | | | | | 338 | | | | 642 | 553 | 695 |
| 18 | | | | | 389 | 181 | | | | 622 | 621 | 511 |
| 19 | | | | | 327 | 281 | | | | 511 | 553 | 238 |
| 20 | | | | | 335 | 319 | | | | 572 | 609 | 483 |
| 21 | | | | | 250 | 336 | | | | 38 | 671 | 583 |
| 22 | | | | | 372 | 307 | | | | 448 | 406 | 356 |
| 23 | | | | | 424 | 332 | | | | 589 | 608 | 701 |
| 24 | | | | | 223 | 337 | | | | 612 | 310 | 572 |
| 25 | | | | | 351 | 334 | | | | 670 | 522 | 397 |
| 26 | | | | | 418 | 307 | | | | 653 | 269 | 481 |
| 27 | | | | | 395 | 272 | | | | 649 | 334 | 588 |
| 28 | | | | | 409 | 200 | | | | 628 | 636 | 437 |
| 29 | | | | | 418 | 168 | | | | 648 | 691 | 660 |
| 30 | | | | | 421 | 194 | | | | 653 | 569 | 584 |
| 31 | | | | | 430 | | | | | 658 | | 588 |

Table 10D: Total Hemispherical Radiation, Fort Dauphin, Madagascar (25S, 47E) (ly day$^{-1}$)

| 1963 | Jan. | Feb. | Mar. | Apr. | May | June | July | Aug. | Sept. | Oct. | Nov. | Dec. |
|------|------|------|------|------|------|------|------|------|-------|------|------|------|
| 1  |  |  |  |      | 1079 | 1008 |      |  |  |  |  |  |
| 2  |  |  |  |      | 1192 | 991  |      |  |  |  |  |  |
| 3  |  |  |  |      | 1160 | 1034 |      |  |  |  |  |  |
| 4  |  |  |  |      | 1081 | 1013 | 876  |  |  |  |  |  |
| 5  |  |  |  |      | 967  | 1063 | 972  |  |  |  |  |  |
| 6  |  |  |  |      | 1091 | 899  | 912  |  |  |  |  |  |
| 7  |  |  |  |      | 1155 | 1017 | 924  |  |  |  |  |  |
| 8  |  |  |  |      | 1183 | 1150 | 978  |  |  |  |  |  |
| 9  |  |  |  |      | 972  | 1114 | 1020 |  |  |  |  |  |
| 10 |  |  |  |      | 1069 | 1105 | 972  |  |  |  |  |  |
| 11 |  |  |  |      | 1152 | 1114 | 1014 |  |  |  |  |  |
| 12 |  |  |  |      | 1150 | 1085 | 864  |  |  |  |  |  |
| 13 |  |  |  |      | 1163 | 1099 | 1032 |  |  |  |  |  |
| 14 |  |  |  |      | 1152 | 1132 | 918  |  |  |  |  |  |
| 15 |  |  |  |      | 1157 | 1052 | 977  |  |  |  |  |  |
| 16 |  |  |  |      | 1149 | 1051 | 1008 |  |  |  |  |  |
| 17 |  |  |  |      | 917  | 986  | 1080 |  |  |  |  |  |
| 18 |  |  |  |      | 1084 | 845  | 1062 |  |  |  |  |  |
| 19 |  |  |  |      | 918  | 848  | 1056 |  |  |  |  |  |
| 20 |  |  |  |      | 1004 | 991  | 1102 |  |  |  |  |  |
| 21 |  |  |  |      | 1101 | 1039 |      |  |  |  |  |  |
| 22 |  |  |  |      | 989  | 1025 |      |  |  |  |  |  |
| 23 |  |  |  |      | 1073 | 1081 |      |  |  |  |  |  |
| 24 |  |  |  | 1031 | 841  | 1031 |      |  |  |  |  |  |
| 25 |  |  |  | 1116 | 990  | 887  |      |  |  |  |  |  |
| 26 |  |  |  | 1165 | 1039 | 887  |      |  |  |  |  |  |
| 27 |  |  |  | 911  | 1069 | 955  |      |  |  |  |  |  |
| 28 |  |  |  | 1145 | 1047 | 1027 |      |  |  |  |  |  |
| 29 |  |  |  | 987  | 1049 | 1041 |      |  |  |  |  |  |
| 30 |  |  |  | 1140 | 1073 | 964  |      |  |  |  |  |  |
| 31 |  |  |  |      | 1064 |      |      |  |  |  |  |  |

Table 10E: Total Hemispherical Radiation, Fort Dauphin, Madagascar (25S, 47E) (ly day$^{-1}$)

| 1961 | Jan. | Feb. | Mar. | Apr. | May | June | July | Aug. | Sept. | Oct. | Nov. | Dec. |
|---|---|---|---|---|---|---|---|---|---|---|---|---|
| 1 | | | | | | | | | | | 1328 | |
| 2 | | | | | | | | | | 1263 | 1011 | 1608 |
| 3 | | | | | | | | | | | 1185 | 1217 |
| 4 | | | | | | | | | | | 1472 | 1462 |
| 5 | | | | | | | | | | | 1333 | 1591 |
| 6 | | | | | | | | | 1247 | | | 1365 |
| 7 | | | | | | | | | 924 | 1079 | | 1292 |
| 8 | | | | | | | | | 1013 | 1148 | 1492 | 1499 |
| 9 | | | | | | | | | 1250 | 1253 | 1447 | 1454 |
| 10 | | | | | | | | | 1303 | 1108 | 1445 | 1528 |
| 11 | | | | | | | | | 1246 | 1268 | 1415 | 1422 |
| 12 | | | | | | | | | 1181 | 1166 | 1436 | 1475 |
| 13 | | | | | | | | | 1032 | 1048 | 1370 | 1501 |
| 14 | | | | | | | | | 1274 | 1276 | 1387 | 1567 |
| 15 | | | | | | | | | 1301 | 1339 | 1431 | 1516 |
| 16 | | | | | | | | | 1303 | 1307 | 1464 | 1498 |
| 17 | | | | | | | | | 1207 | 1396 | 1439 | 1508 |
| 18 | | | | | | | | | 1303 | 1362 | 1439 | 1306 |
| 19 | | | | | | | | | 1220 | 1404 | 1375 | 1507 |
| 20 | | | | | | | | | 1270 | 1428 | 1399 | 1177 |
| 21 | | | | | | | | | 959 | 1367 | 1364 | 1112 |
| 22 | | | | | | | | | 1171 | 1286 | 1340 | 1273 |
| 23 | | | | | | | | | 1102 | 1381 | | 1373 |
| 24 | | | | | | | | | 1019 | 1271 | | 1171 |
| 25 | | | | | | | | | 1310 | 1308 | | 959 |
| 26 | | | | | | | | | 1274 | 1331 | | 1262 |
| 27 | | | | | | | | | 1158 | 1306 | 1269 | 1441 |
| 28 | | | | | | | | | 1345 | 1315 | 1267 | 1504 |
| 29 | | | | | | | | | | 1230 | | 1487 |
| 30 | | | | | | | | | | 1250 | | 1471 |
| 31 | | | | | | | | | | 1060 | | 1397 |

Table 10F: Total Hemispherical Radiation, Fort Dauphin, Madagascar (25S, 47E) (ly day$^{-1}$)

| 1965 | Jan. | Feb. | Mar. | Apr. | May | June | July | Aug. | Sept. | Oct. | Nov. | Dec. |
|------|------|------|------|------|-----|------|------|------|-------|------|------|------|
| 1 | | | | | | | | | | | | |
| 2 | | | | | | | | | | | | |
| 3 | | | | | | | | | | | | |
| 4 | | | | | | | | | | | | |
| 5 | | | | | | | | | | | | |
| 6 | | 1443 | | | | | | | | | | |
| 7 | | 1246 | | | | | | | | | | |
| 8 | | | | | | | | | | | | |
| 9 | | | | | | | | | | | | |
| 10 | | 1658 | | | | | | | | | | |
| 11 | | 1537 | | | | | | | | | | |
| 12 | | 1488 | | | | | | | | | | |
| 13 | | 1452 | | | | | | | | | | |
| 14 | | 1444 | | | | | | | | | | |
| 15 | | 1446 | | | | | | | | | | |
| 16 | | | | | | | | | | | | |
| 17 | | | | | | | | | | | | |
| 18 | | | | | | | | | | | | |
| 19 | | | | | | | | | | | | |
| 20 | | | | | | | | | | | | |
| 21 | | | | | | | | | | | | |
| 22 | | | | | | | | | | | | |
| 23 | | | | | | | | | | | | |
| 24 | | | | | | | | | | | | |
| 25 | | | | | | | | | | | | |
| 26 | | | | | | | | | | | | |
| 27 | | | | | | | | | | | | |
| 28 | | | | | | | | | | | | |
| 29 | | | | | | | | | | | | |
| 30 | | | | | | | | | | | | |
| 31 | | | | | | | | | | | | |

Table 11A: Solar Radiation, Plaisance, Mauritius Island (20S, 57E) (ly day$^{-1}$)

| 1963 | Jan | Feb | Mar | Apr | May | June | July | Aug | Sept | Oct | Nov. | Dec. |
|------|-----|-----|-----|-----|-----|------|------|-----|------|-----|------|------|
| 1 | | | | | | | 142 | 415 | 295 | 338 | 114 | 629 |
| 2 | | | | | | | 272 | 259 | 423 | 503 | 565 | 705 |
| 3 | | | | | | | 209 | 287 | 463 | 543 | 602 | 667 |
| 4 | | | | | | | 337 | 406 | | 538 | 500 | 590 |
| 5 | | | | | | | 289 | 190 | 346 | 460 | 337 | 598 |
| 6 | | | | | | | 252 | 364 | 444 | 358 | 563 | 670 |
| 7 | | | | | | | 322 | 371 | 353 | 344 | 498 | 656 |
| 8 | | | | | | | 175 | 229 | 430 | 519 | 417 | 352 |
| 9 | | | | | | | 249 | 159 | 480 | 605 | 410 | 646 |
| 10 | | | | | | | 276 | 442 | 450 | 421 | 437 | 200 |
| 11 | | | | | | | 286 | 214 | 420 | 279 | 525 | 422 |
| 12 | | | | | | | 251 | 416 | 388 | 588 | 573 | 579 |
| 13 | | | | | | | 307 | 316 | 419 | 518 | 567 | 626 |
| 14 | | | | | | | 348 | 459 | 412 | 257 | 235 | 640 |
| 15 | | | | | | | 371 | 476 | 295 | 355 | 289 | 571 |
| 16 | | | | | | | 374 | 463 | 428 | 495 | 429 | 610 |
| 17 | | | | | | | 383 | 384 | 344 | 425 | 446 | 579 |
| 18 | | | | | | | 326 | 403 | 406 | 176 | 617 | 709 |
| 19 | | | | | | | 219 | 295 | 523 | 441 | 446 | 660 |
| 20 | | | | | | | 346 | 357 | 331 | 343 | 276 | 289 |
| 21 | | | | | | | 338 | 428 | 432 | 660 | 624 | 649 |
| 22 | | | | | | | 361 | 422 | 366 | 555 | 578 | 494 |
| 23 | | | | | | | 381 | 450 | 565 | 467 | 427 | 593 |
| 24 | | | | | | | 246 | 345 | 488 | 460 | 255 | 601 |
| 25 | | | | | | | 384 | 291 | 471 | 373 | 513 | 529 |
| 26 | | | | | | | 285 | 256 | 369 | 478 | 476 | 637 |
| 27 | | | | | | | 221 | 401 | 591 | 277 | 298 | 511 |
| 28 | | | | | | | 306 | | 495 | 495 | 295 | 676 |
| 29 | | | | | | | 301 | | | | 692 | 680 |
| 30 | | | | | | | | | | | | 612 |
| 31 | | | | | | | | | | | | |

Table 11B: Solar Radiation, Plaisance, Mauritius Island (20S, 57E) (ly day$^{-1}$)

| 1964 | Jan. | Feb. | Mar. | Apr. | May | June | July | Aug. | Sept. | Oct. | Nov. | Dec. |
|------|------|------|------|------|-----|------|------|------|-------|------|------|------|
| 1  | 648 |     | 590 | 375 |     |     | 345 |     | 249 | 381 | 540 | 658 |
| 2  | 358 | 215 | 537 | 367 |     |     | 222 | 374 | 262 | 428 | 393 | 726 |
| 3  | 581 | 604 | 479 | 474 |     |     | 311 | 370 | 283 | 580 | 283 | 590 |
| 4  | 591 | 687 |     |     | 67  | 250 | 328 | 331 | 350 | 390 | 621 | 509 |
| 5  | 618 | 534 | 245 |     | 267 | 313 | 310 | 364 | 379 | 405 | 446 | 511 |
| 6  | 536 | 405 | 308 |     | 205 | 199 | 404 | 326 | 392 | 341 | 511 | 440 |
| 7  | 554 |     | 546 | 379 | 343 | 323 | 347 | 319 | 420 | 436 | 518 | 702 |
| 8  | 556 | 396 | 493 | 386 | 262 | 205 | 367 | 403 | 500 | 355 | 540 | 704 |
| 9  | 408 | 254 | 293 | 477 | 355 | 128 | 305 | 340 | 260 | 333 | 493 | 180 |
| 10 | 642 | 570 | 240 | 397 | 394 | 311 | 328 | 413 | 391 | 385 | 505 | 474 |
| 11 | 615 | 454 | 467 | 378 | 212 | 286 | 367 |     | 315 | 446 | 702 | 512 |
| 12 | 543 | 519 | 413 | 278 | 326 | 275 | 329 |     | 305 | 538 | 689 | 527 |
| 13 | 429 | 595 | 419 | 460 | 246 | 338 | 224 | 281 | 456 | 523 | 746 | 516 |
| 14 | 510 | 458 | 481 | 416 | 228 | 241 | 243 | 474 | 389 | 599 | 678 | 595 |
| 15 | 478 | 583 | 368 | 340 | 403 | 308 | 172 | 292 | 461 | 594 | 499 | 370 |
| 16 | 550 | 423 | 436 | 347 | 299 | 281 | 379 | 405 | 450 | 643 | 730 | 643 |
| 17 |     | 571 | 505 | 328 | 299 | 180 | 363 | 348 | 479 | 358 | 602 | 564 |
| 18 |     | 496 | 451 | 437 | 425 | 295 |     | 457 | 428 | 446 | 704 | 589 |
| 19 |     | 440 | 380 | 340 | 246 | 349 | 328 | 320 | 387 | 631 | 710 | 713 |
| 20 | 280 |     | 268 | 362 | 336 | 392 | 406 | 419 | 535 | 610 | 632 | 739 |
| 21 | 604 | 320 | 504 | 346 | 232 | 317 | 401 | 266 | 389 | 426 | 676 | 748 |
| 22 | 674 | 604 | 468 | 410 | 387 | 322 | 239 | 445 | 392 | 489 | 629 | 700 |
| 23 | 527 | 535 | 542 | 350 | 343 | 447 | 322 | 299 | 539 | 660 | 736 | 634 |
| 24 | 662 | 545 | 368 | 187 | 379 | 240 | 208 | 396 | 512 | 665 | 622 | 728 |
| 25 | 429 | 341 | 506 | 239 | 398 | 331 | 300 | 370 | 413 | 456 | 718 | 436 |
| 26 | 311 | 219 | 308 | 205 | 377 | 338 | 403 | 326 | 447 | 344 | 724 | 716 |
| 27 | 663 | 145 | 317 | 319 | 409 | 245 | 362 | 366 | 592 | 415 | 742 | 324 |
| 28 | 570 | 190 | 433 | 422 | 310 | 383 | 223 | 374 | 469 | 516 | 715 | 446 |
| 29 | 637 | 298 | 282 | 410 | 269 | 299 | 390 | 311 | 596 | 304 | 633 | 479 |
| 30 | 350 |     |     | 386 |     | 338 | 285 |     |     | 689 | 575 | 541 |
| 31 | 359 |     | 445 |     |     |     | 288 | 449 |     | 700 |     | 512 |

Table 11C: Solar Radiation, Plaisance, Mauritius Island (20S, 57E) (ly day$^{-1}$)

| 1965 | Jan. | Feb. | Mar. | Apr. | May | June | July | Aug. | Sept. | Oct. | Nov. | Dec. |
|---|---|---|---|---|---|---|---|---|---|---|---|---|
| 1 | | | | | 481 | | | 388 | 293 | 608 | 565 | |
| 2 | | | | | 156 | | | 326 | 371 | 614 | 400 | 631 |
| 3 | | | | | 66 | | | | 388 | 433 | 607 | 557 |
| 4 | | | | | 182 | | 304 | 374 | | | 469 | 557 |
| 5 | | | | | 490 | | 93 | 401 | | 301 | 297 | 669 |
| 6 | | | | | 607 | | 215 | 220 | | 425 | 599 | 670 |
| 7 | | | | | 160 | | 248 | | 452 | 466 | 640 | 666 |
| 8 | | | | | 239 | | 229 | | 381 | 409 | 598 | 630 |
| 9 | | | | | 425 | | 247 | 430 | 338 | 500 | 453 | 688 |
| 10 | | | | | 374 | 313 | 239 | 406 | 532 | 613 | 601 | 436 |
| 11 | | | | | | 332 | 379 | 248 | 323 | 349 | 614 | 637 |
| 12 | | | | | | 226 | 227 | 286 | 457 | 549 | 649 | 674 |
| 13 | | | | | 447 | 142 | 276 | 382 | 440 | 553 | 628 | |
| 14 | | | | | 430 | 308 | 233 | 310 | 487 | 598 | 555 | 533 |
| 15 | | | | | 327 | 152 | 202 | 411 | 400 | 555 | 489 | 658 |
| 16 | | | | | | 224 | 347 | 154 | 482 | | 472 | 631 |
| 17 | | | | | | | 150 | 252 | 513 | 622 | 585 | 700 |
| 18 | | | | | | | 205 | | 514 | 662 | 708 | 680 |
| 19 | | | | | | | 313 | 369 | 410 | 455 | 728 | 679 |
| 20 | | | | | | | 207 | 407 | 511 | 485 | 450 | 538 |
| 21 | | | | | | | 241 | 283 | 520 | 671 | 405 | 363 |
| 22 | | | | | | | 322 | 431 | 358 | 668 | 616 | 641 |
| 23 | | | | | | 359 | 193 | 418 | 470 | 579 | 593 | 503 |
| 24 | | | | | | 367 | 295 | 252 | 451 | 457 | 760 | 584 |
| 25 | | | | | | 245 | 368 | 424 | 327 | 389 | 481 | 409 |
| 26 | | | | 418 | | 376 | 325 | 289 | 428 | 550 | 584 | 460 |
| 27 | | | | 358 | | 312 | 178 | 452 | 134 | 479 | 639 | 562 |
| 28 | | | | 365 | | 340 | 275 | 387 | 370 | 544 | 452 | 569 |
| 29 | | | | 187 | | 224 | 322 | 412 | 179 | 683 | 701 | 601 |
| 30 | | | | 558 | | 181 | 251 | 182 | 488 | 580 | | 468 |
| 31 | | | | | | | 285 | 176 | | 428 | | 406 |

Table 11D: Total Hemispherical Radiation, Plaisance, Mauritius Island (20S, 57E) (ly day$^{-1}$)

| 1963 | Jan. | Feb. | Mar. | Apr. | May | June | July | Aug. | Sept. | Oct. | Nov. | Dec. |
|---|---|---|---|---|---|---|---|---|---|---|---|---|
| 1 | | | | | | | | | 1223 | | | |
| 2 | | | | | | | | 1121 | 1360 | | | |
| 3 | | | | | | | 1063 | 1144 | 1397 | | | |
| 4 | | | | | | | 1196 | 1200 | | | | |
| 5 | | | | | | | 1106 | 1030 | 1191 | | | |
| 6 | | | | | | | 1102 | 1195 | 1254 | | | |
| 7 | | | | | | | 1177 | 1186 | 1907 | | | |
| 8 | | | | | | | 1052 | 1127 | 1207 | | | |
| 9 | | | | | | | 1095 | 1063 | 1325 | | | |
| 10 | | | | | | | 1077 | 1318 | 1262 | | | |
| 11 | | | | | | | 1132 | 1125 | 1239 | | | |
| 12 | | | | | | | 1162 | 1279 | 1194 | | | |
| 13 | | | | | | | 1139 | 1203 | 1268 | | | |
| 14 | | | | | | | 1147 | 1244 | 1266 | | | |
| 15 | | | | | | | 1220 | 1363 | 1120 | | | |
| 16 | | | | | | | 1227 | 1449 | 1262 | | | |
| 17 | | | | | | | 1226 | 1313 | 1191 | | | |
| 18 | | | | | | | 1153 | 1362 | 1321 | | | |
| 19 | | | | | | | 1008 | 1218 | 1337 | | | |
| 20 | | | | | | | 1237 | 1316 | 1267 | | | |
| 21 | | | | | | | 1205 | 1407 | 1217 | | | |
| 22 | | | | | | | 1157 | 1398 | 1152 | | | |
| 23 | | | | | | | 1223 | 1335 | 1316 | | | |
| 24 | | | | | | | 1071 | 1261 | 1268 | | | |
| 25 | | | | | | | 1283 | 1178 | 1315 | | | |
| 26 | | | | | | | 1235 | 1172 | 1246 | | | |
| 27 | | | | | | | 1183 | 1320 | 1347 | | | |
| 28 | | | | | | | 1129 | 1316 | | | | |
| 29 | | | | | | | 1121 | | | | | |
| 30 | | | | | | | | | | | | |
| 31 | | | | | | | | | | | | |

**Table 11E: Total Hemispherical Radiation, Plaisance, Mauritius Island (20S, 57E) (ly day$^{-1}$)**

| 1904 | Jan. | Feb. | Mar. | Apr. | May | June | July | Aug. | Sept. | Oct. | Nov. | Dec. |
|---|---|---|---|---|---|---|---|---|---|---|---|---|
| 1 | | | | 1231 | | | | | 1068 | 1260 | 1410 | 1546 |
| 2 | | | 1582 | 1255 | | | | 1238 | 1137 | 1331 | 1169 | 1600 |
| 3 | | | 1453 | 1264 | | | | 1089 | 1107 | 1444 | 1102 | 1406 |
| 4 | | | 1532 | | | | | 1123 | 1231 | 1253 | 1483 | 1454 |
| 5 | | | | 1320 | | | | 1157 | 1275 | 1311 | 1274 | 1409 |
| 6 | | | | 1200 | | | | 1118 | 1237 | 1189 | 1420 | 1428 |
| 7 | | | | 1146 | | | | 1111 | 1226 | 1228 | 1360 | 1591 |
| 8 | | 1456 | 1448 | 1318 | | | | 1195 | 1338 | 1159 | 1374 | 1604 |
| 9 | | 1325 | 1219 | 1397 | | | | 1132 | 1036 | 1201 | 1351 | |
| 10 | | 1601 | | 1298 | | | | 1205 | 1216 | 1265 | 1354 | 1398 |
| 11 | | 1426 | 1422 | 1364 | | | | | 1100 | 1276 | 1553 | 1390 |
| 12 | | 1475 | 1298 | 1224 | | | | | 1127 | 1395 | 1528 | 1190 |
| 13 | | 1530 | 1363 | 1281 | | | | 1073 | 1275 | 1330 | 1576 | |
| 14 | | 1405 | 1523 | 1290 | | | | 1266 | 1192 | 1381 | 1487 | |
| 15 | | 1531 | 1327 | 1195 | | | | 1084 | 1255 | 1342 | 1409 | |
| 16 | | 1388 | 1357 | 1184 | | | | 1197 | 1235 | 1497 | 1565 | 1427 |
| 17 | | 1507 | 1398 | 1214 | | | | 1140 | 1286 | 1165 | 1445 | 1499 |
| 18 | | 1397 | 1515 | 1220 | | | | 1321 | 1278 | 1263 | 1559 | 1472 |
| 19 | | 1462 | 1321 | 1177 | | | | 1112 | 1234 | 1457 | 1531 | 1610 |
| 20 | | | 1186 | | | | | 1211 | 1330 | 1453 | 1486 | 1572 |
| 21 | | | 1447 | | | | | 1058 | 1239 | 1264 | 1519 | 1572 |
| 22 | | 1492 | 1354 | | | | | 1237 | 1217 | 1332 | 1479 | 1587 |
| 23 | | 1439 | 1495 | | | | | 1091 | 1339 | 1496 | 1574 | 1521 |
| 24 | | 1405 | 1354 | | | | | 1188 | 1369 | 1507 | 1465 | 1590 |
| 25 | | 1355 | 1402 | | | | | 1162 | 1316 | 1288 | 1590 | |
| 26 | | | 1367 | | | | | 1118 | 1390 | 1198 | 1623 | 1588 |
| 27 | | | 1435 | | | | | 1158 | 1529 | 1278 | 1654 | |
| 28 | | | 1393 | | | | | 1166 | 1366 | 1353 | 1601 | |
| 29 | | | 1279 | | | | | 1103 | 1438 | 1190 | 1486 | |
| 30 | | | | | | | | | | 1581 | 1499 | |
| 31 | | | 1477 | | | | | 1259 | | 1556 | | |

Table 11F: Total Hemispherical Radiation, Plaisance, Mauritius Island (20S, 57E) (ly day$^{-1}$)

| 1965 | Jan. | Feb. | Mar. | Apr. | May | June | July | Aug. | Sept. | Oct. | Nov. | Dec. |
|---|---|---|---|---|---|---|---|---|---|---|---|---|
| 1 | | | | | | | | 1137 | | | | |
| 2 | | | | | | | | | | | | |
| 3 | | | | | | | | 1220 | | | | |
| 4 | | | | | | | 1160 | 1205 | | | | |
| 5 | | | | | | | 1193 | 1174 | | | | |
| 6 | | | | | | | 1197 | 1008 | | | | |
| 7 | | | | | | | 1097 | | | | | |
| 8 | | | | | | | 1076 | | | | | |
| 9 | | | | | | | 1127 | 1182 | | | | |
| 10 | | | | | | 1111 | 1068 | 1144 | | | | |
| 11 | | | | | | 1133 | 1180 | 1048 | | | | |
| 12 | | | | | | 1087 | 1084 | | | | | |
| 13 | | | | | | 1046 | 1133 | | | | | |
| 14 | | | | | | 1243 | 1138 | | | | | |
| 15 | | | | | | 1025 | 987 | | | | | |
| 16 | | | | | | 1112 | 1126 | | | | | |
| 17 | | | | | | 1144 | 964 | | | | | |
| 18 | | | | | | 1246 | 960 | | | | | |
| 19 | | | | | | 1143 | 1077 | | | | | |
| 20 | | | | | | 1036 | 948 | | | | | |
| 21 | | | | | | 1070 | 1001 | | | | | |
| 22 | | | | | | 1126 | 1026 | | | | | |
| 23 | | | | | | 1149 | 960 | | | | | |
| 24 | | | | | | 1101 | 1105 | | | | | |
| 25 | | | | | | 1117 | 1229 | | | | | |
| 26 | | | | | | 1183 | 1184 | | | | | |
| 27 | | | | | | 1150 | 983 | | | | | |
| 28 | | | | | | 1169 | 1055 | | | | | |
| 29 | | | | | | 1009 | 1105 | | | | | |
| 30 | | | | | | 1019 | 1019 | | | | | |
| 31 | | | | | | | 1057 | | | | | |

**Table 12A: Solar Radiation, Mahé Island, Seychelles Islands (5S, 55E) (ly day$^{-1}$)**

| 1963 | Jan. | Feb. | Mar. | Apr. | May | June | July | Aug. | Sept. | Oct. | Nov. | Dec. |
|---|---|---|---|---|---|---|---|---|---|---|---|---|
| 1 | | | | | | | | | | | 590 | 608 |
| 2 | | | | | | | | | | | 559 | 410 |
| 3 | | | | | | | | | | | 555 | 334 |
| 4 | | | | | | | | | | | 655 | 459 |
| 5 | | | | | | | | | | | 557 | 573 |
| 6 | | | | | | | | | | | 510 | 490 |
| 7 | | | | | | | | | | | 631 | 584 |
| 8 | | | | | | | | | | | 222 | 583 |
| 9 | | | | | | | | | | | 172 | 377 |
| 10 | | | | | | | | | | | 437 | 596 |
| 11 | | | | | | | | | | | 586 | 663 |
| 12 | | | | | | | | | | | 606 | 373 |
| 13 | | | | | | | | | | | 664 | 379 |
| 14 | | | | | | | | | | | 618 | 443 |
| 15 | | | | | | | | | | | 467 | 356 |
| 16 | | | | | | | | | | | 511 | 500 |
| 17 | | | | | | | | | | | 370 | 495 |
| 18 | | | | | | | | | | | 390 | 598 |
| 19 | | | | | | | | | | | 387 | 468 |
| 20 | | | | | | | | | | | 566 | 371 |
| 21 | | | | | | | | | | | 246 | 346 |
| 22 | | | | | | | | | | | 199 | 151 |
| 23 | | | | | | | | | | | | |
| 24 | | | | | | | | | | | 463 | 410 |
| 25 | | | | | | | | | | | 248 | 602 |
| 26 | | | | | | | | | | | 639 | 538 |
| 27 | | | | | | | | | | 532 | 506 | |
| 28 | | | | | | | | | | | 650 | |
| 29 | | | | | | | | | | 464 | 380 | |
| 30 | | | | | | | | | | 517 | 598 | |
| 31 | | | | | | | | | | 449 | 668 | |

Table 12B: Solar Radiation, Mahé Island, Seychelles Islands (5S, 55E) (ly day⁻¹)

| 1964 | Jan. | Feb. | Mar. | Apr. | May | June | July | Aug. | Sept. | Oct. | Nov. | Dec. |
|---|---|---|---|---|---|---|---|---|---|---|---|---|
| 1 | 564 | 73 | 526 | 560 | 607 | 454 | 521 | 404 | 496 | 362 | 292 | 332 |
| 2 | 585 | 298 | 656 | 613 | 532 | 366 | 543 | 481 | 507 | 565 | 625 | 449 |
| 3 | 371 | 355 | 529 | 436 | 562 | 432 | 345 | | 494 | 683 | 398 | 630 |
| 4 | 599 | 511 | 522 | 637 | 536 | 448 | 516 | 422 | 276 | 516 | 715 | 668 |
| 5 | 487 | 571 | 539 | 440 | 582 | 510 | 442 | 515 | 560 | 445 | | 680 |
| 6 | 518 | 685 | 548 | 335 | 521 | 534 | 437 | 337 | 534 | 406 | 665 | 559 |
| 7 | 385 | 650 | 675 | 603 | 575 | 341 | 387 | 197 | 384 | 466 | 651 | 280 |
| 8 | 377 | 690 | 627 | 527 | 495 | 495 | 340 | 323 | 543 | 617 | 534 | 214 |
| 9 | 548 | 674 | 193 | 635 | 527 | 488 | 463 | 505 | 374 | 574 | 691 | 586 |
| 10 | 269 | 567 | 251 | 613 | 572 | 505 | 266 | 538 | 184 | 581 | 610 | 591 |
| 11 | 315 | 463 | 650 | 651 | 344 | 500 | 326 | 598 | 51 | 684 | 509 | 527 |
| 12 | 157 | 299 | 456 | 179 | 425 | 429 | 349 | 571 | 507 | 678 | | 540 |
| 13 | 495 | 551 | 377 | 87 | 478 | 383 | 449 | 599 | 593 | | 685 | 656 |
| 14 | 419 | 340 | 359 | 496 | 449 | 238 | 320 | 588 | 637 | 581 | 507 | 511 |
| 15 | 343 | 637 | 614 | 503 | 529 | 379 | 423 | 571 | 607 | 637 | 331 | |
| 16 | 167 | 325 | 326 | 548 | 226 | 415 | 494 | 558 | 544 | 393 | 336 | 680 |
| 17 | 293 | 178 | 429 | 616 | 253 | 483 | 448 | 476 | 508 | 654 | 344 | 614 |
| 18 | 412 | 299 | 329 | 377 | 202 | 340 | 416 | 602 | 591 | 201 | 376 | |
| 19 | 427 | 232 | 459 | 464 | 378 | 347 | 433 | 601 | 605 | 299 | 535 | 480 |
| 20 | 617 | 329 | 543 | 536 | 473 | 181 | 226 | 249 | 614 | 386 | 326 | 313 |
| 21 | 562 | 432 | 411 | 289 | 490 | 343 | 64 | 92 | 353 | 491 | 603 | 209 |
| 22 | 206 | 265 | 577 | 472 | 547 | 508 | 134 | 559 | 254 | 206 | 360 | 360 |
| 23 | 157 | 622 | 597 | 137 | 428 | 478 | 440 | 406 | 565 | 381 | 573 | 119 |
| 24 | 509 | 671 | 680 | 293 | 439 | 401 | 310 | 508 | 605 | 170 | 662 | 194 |
| 25 | 115 | 398 | 620 | 307 | 323 | 427 | 546 | 435 | 634 | 641 | 626 | 108 |
| 26 | 526 | 602 | 616 | 558 | 63 | 371 | 505 | 393 | 628 | 506 | 462 | 347 |
| 27 | 347 | 560 | 425 | 605 | 518 | 425 | 390 | 233 | 577 | 647 | 353 | 128 |
| 28 | 50 | | 413 | | 569 | 364 | 434 | 420 | 613 | 337 | 335 | 103 |
| 29 | 181 | | 526 | | 538 | 292 | 418 | 400 | 365 | 155 | 425 | 530 |
| 30 | 433 | | 575 | | 473 | 295 | | | | 341 | 412 | 437 |
| 31 | 541 | | 529 | | | | | | | 489 | | |

**Table 12C: Solar Radiation, Mahé Island, Seychelles Islands (5S, 55E) (ly day$^{-1}$)**

| 1965 | Jan. | Feb. | Mar. | Apr. | May | June | July | Aug | Sept. | Oct. | Nov. | Dec. |
|---|---|---|---|---|---|---|---|---|---|---|---|---|
| 1 | 338 | 496 | 403 | 526 | | | | | | | | |
| 2 | 319 | 569 | 575 | 675 | | | | | | | | |
| 3 | 552 | | 485 | 517 | | | | | | | | |
| 4 | 675 | 338 | 391 | 506 | | | | | | | | |
| 5 | 611 | 392 | 311 | 602 | | | | | | | | |
| 6 | 443 | 419 | 313 | 416 | | | | | | | | |
| 7 | 147 | 488 | 410 | | | | | | | | | |
| 8 | 312 | 470 | 694 | | | | | | | | | |
| 9 | 421 | 506 | 623 | | | | | | | | | |
| 10 | 553 | 460 | 623 | | | | | | | | | |
| 11 | | 208 | 703 | | | | | | | | | |
| 12 | 671 | 265 | 564 | | | | | | | | | |
| 13 | 517 | 42 | 669 | | | | | | | | | |
| 14 | 530 | 326 | 661 | | | | | | | | | |
| 15 | 635 | 243 | 623 | | | | | | | | | |
| 16 | 651 | 384 | 375 | | | | | | | | | |
| 17 | 643 | 539 | 550 | | | | | | | | | |
| 18 | 610 | 404 | 706 | | | | | | | | | |
| 19 | 647 | 167 | 472 | | | | | | | | | |
| 20 | 632 | 113 | 467 | | | | | | | | | |
| 21 | 582 | 89 | 350 | | | | | | | | | |
| 22 | 470 | 224 | 628 | | | | | | | | | |
| 23 | 324 | 328 | 671 | | | | | | | | | |
| 24 | 549 | 49 | 655 | | | | | | | | | |
| 25 | 332 | 401 | 685 | | | | | | | | | |
| 26 | 445 | 332 | 665 | | | | | | | | | |
| 27 | 204 | 382 | 443 | | | | | | | | | |
| 28 | 408 | 323 | 455 | | | | | | | | | |
| 29 | 271 | 323 | 374 | | | | | | | | | |
| 30 | 221 | | 294 | | | | | | | | | |
| 31 | 202 | | 353 | | | | | | | | | |

# Notes and References

1. Professor of Meteorology, University of Michigan.

2. Research Associate, Department of Meteorology and Oceanography, University of Michigan.

3. Budyko, M. I. (ed.), 1963: *Atlas teplovogo balansa zemnogo shara* (Atlas of the heat balance of the earth). Moscow, 69 pp.

4. Mani, A., O. Chacko, V. Krishnamurthy, and V. Desikan, 1967: Distribution of global and net radiation over the Indian Ocean and its environments. *Arch. Meteor. Bioklim.*, B, **15**, 82-98.

5. Complete information on the pyrheliometer is available in Bulletin No. 2, August 1964, published by the Eppley Laboratory, Inc., Newport, R.I., and its characteristics have been described by

   Kimball, H. H., and H. E. Hobbs, 1923: A new form of thermoelectric recording pyrheliometer. *Mon. Wea. Rev.*, **51**, 239–242.

   Hand, I. F., 1946: Pyrheliometers and pyrheliometric measurements. *U.S. Dept. Commerce, Weather Bur. Circ. Q.*, 55 pp.

   Fuquay, D., and K. Buettner, 1957: Laboratory investigation of some characteristics of the Eppley Pyrheliometer. *Trans. Amer. Geophys. Union*, **38**, 38–43.

6. Gier, J. T., and R. V. Dunkle, 1951: Total hemispherical radiometers. *Proc. Amer. Inst. Electr. Engrs.*, **70**, 339–343.

7. Portman, D. J., and F. Dias, 1959: Influence of wind and angle of incident radiation on the performance of a Beckman and Whitley total hemispherical radiometer. *University of Michigan Res. Inst. Rept.* 2715–1–F, 38 pp.

8. MacKay, K. P., Jr., 1965: An improved total hemispherical radiometer. *J. App. Meteor.*, **4**, 112–115.

9. Bolsenga, S. J., 1964: Daily sums of global radiation for cloudless skies. *U.S. Army Materiel Command Cold Regions Research and Engineering Laboratory Res. Rept.* No. 160, 124 pp.

10. Kimball, H. H., 1927: Measurements of solar radiation intensity and determination of its depletion by the atmosphere with bibliography of pyrheliometric measurements. *Mon. Wea. Rev.*, **55**, 155–169.

    _____, 1930: Measurements of solar radiation intensity and determinations of its depletion in the atmosphere. *Mon. Wea. Rev.*, **58**, 43–52.

11. Drummond, A. J., and E. Vowinckel, 1957: The distribution of solar radiation throughout southern Africa. *J. Meteor.*, **14**, 343–353.

12. Flowers, E. C., and H. J. Viebrock, 1965: Solar radiation: an anomalous decrease of direct solar radiation. *Science*, **148**, 493–494.

13. Budyko, M. I., and Z. I. Pivovarova, 1967: Vlianie vulcanicheskykh izverzhenii na prychodyashchuyu k poverkhnosty zemly solnechnuyu radiatsiyu (The effect of volcanic eruptions on incoming solar radiation at the earth's surface). *Meteorologiia i Gidrologiia*, **10**, 3–7.

14. Kondrat'yev, K. Ya., G. A. Nicolsky, I. Ya. Badinov, and S. D. Andreev, 1967: Direct solar radiation up to 30 km and stratification of attenuation components in the stratosphere. *App. Opt.*, **6**, 197–206.

15. Mani, A., and O. Chacko, 1963: Measurements of solar radiation and atmospheric turbidity with Angstrom pyrheliometers at Poona and Delhi during the IGY. *Ind. J. Meteor. Geophys.*, **14**, 271–279.

16. Kondrat'yev, K. Ya., 1965: *Actinometry*. Translated by National Aeronautics and Space Administration NASATT F–9712, 675 pp.

17. Ta-k'ang, I-hsien, and Chien-sui, 1963: *Characteristics of the distribution of total radiation in China*. U.S. Dept. Commerce, JPRS 36, 303, TT66–32735, p. 11–36.

18. Berlyand, T. C., 1960: Metodika Klimatologicheskikh raschetov summarnoi radiatsii (Method of climatological estimation of global radiation). *Meteorologiia i Gidrologiia*, **6**, 9–12.

19. Ter-Markaryants, N. E., 1960: On the average daily values of the sea's albedo. *Trans. (Trudy) Main Geophys. Observ.*, Issue 100.

20. Burt, W. V., 1954: Albedo over wind roughened water. *J. Meteor.*, **2**, 283–290.

21. Degtyarev, G. M., A. A. Sorokin, and Yu. A. Men'shov, 1964: Radiation reflected from the surface of the open sea. Translated by Stephen B. Dresner. *Izv. Geophys. Ser.*, No. 10, 953–956.

22. Pivovarov, A. A., E. P. Anisimova, and A. N. Erikova, 1965: Diurnal course of the albedo and of solar radiation penetrating the sea. *Izv. Atmos. Ocean Physics*, **1**, 713–715.

23. Beard, J. T., and J. A. Wiebelt, 1966: Reflectance of a water wave surface as related to evaporation suppression. *J. Geophys. Res.*, **71**, 3843–3847.

24. Koberg, G. E., 1964: Methods to compute long-wave radiation from the atmosphere and reflected solar radiation from a water surface. *Geol. Surv. Prof. Paper*, 272–F. Washington, D.C., U.S. Govt. Printing Office.

25. Anderson, E. R., 1954: Energy-budget studies, p. 70–119. *In* Water-loss Investigations: Lake Hefner Studies. *Geol. Surv. Prof. Paper* 269. Washington, D.C., U.S. Govt. Printing Office.

26. Swinbank, W. C., 1963: Long-wave radiation from clear skies. *Quart. J. Royal Meteor. Soc.*, **89**, 339–348.

27. Hinzpeter, H., 1967: Results of radiation balance measurements during the cruise between Suez and Aden II of the Indian Ocean Expedition 1964–1965. *"Meteor" Forschungsergebnisse*, B, No. 1, 1–13.

28. Winston, J. S., 1967: Planetary-scale characteristics of monthly mean long-wave radiation and albedo and some year-to-year variations. *Mon. Wea. Rev.*, **95**, 235–255.

29. Adem, J., 1967: On the relations between outgoing long-wave radiation, albedo, and cloudiness. *Mon. Wea. Rev.*, **95**, 257–260.

30. Huschke, R. E., 1959: *Glossary of Meteorology*. Boston, American Meteorological Society, 638 pp.

31. Tetens, O., 1930: Uber einige meteorologische Begriffe. *Z. Geophys.*, **6**, 297–309.